Our Self Assembling Universe

An OSAU-2 C&I&L&E=mc² AWTbook™

H. Frank Gaertner

WORKBOOK PRESS LLC
187 E Warm Springs Rd,
Suite B285, Las Vegas, NV 89119, USA

Website: https://workbookpress.com/
Hotline: 1-888-818-4856
Email: admin@workbookpress.com

Ordering Information:
Quantity sales. Special discounts are available on quantity purchases by corporations, associations, and others.
For details, contact the publisher at the address above.

ISBN-13: 978-1-957618-30-2 (Paperback Version)
978-1-957618-31-9 (Digital Version)

REV. DATE: 27/01/2022

Contents

Dedication

This book is dedicated to all grade-school science teachers, my six grandchildren, two of which are already budding physicists, and to the more than 300 8th-graders of El Cajon Montgomery Middle School who participated in my May, 2017, ping-pong-ball, nuclear-physics workshop where they assembled on their own, memorable, take-home models of the alpha-particle (helium-4), beryllium-8, carbon-12, oxygen-16 and neon-20.

Forward

This revised edition of *Our Self-Assembling Universe, OSAU-2,* eliminates errors that were present in the first edition, and adds content to make it what I've dubbed to be called an AWTbook™, a book that may well be the first of many such books that become the best teaching-tools ever devised. I have already witnessed the profound, positive effects gained when I used the first version of this book to teach 8th-graders aspects of astrophysics, chemistry and nuclear physics. My presentation used the origin of Our Universe as a unique starting-point that made my introduction to the physical sciences interesting, exciting and accessible. I know my approach worked well with 8th-graders but what about even younger children? For example, what if children knew practically from day-one that atoms assemble themselves from electrons, protons and neutrons? What if they knew a little about quantum mechanics, how our sun works, and how all main sequence stars assemble the nucleus of the all-important alpha-particle, helium-4? Sounds ridiculous doesn't it? However, armed with ping-pong balls and short, inspiring videos, I believe students of any age may be able to easily understand as they assemble for themselves tangible representatives of the intangible-and-too-small-to-imagine world of our protons and neutrons. And with the challenging fun I've witnessed with young people trying to work with fabric tape for the first time, this type of early introduction to fusion and its role in our universe is so fun and easy to do that it might even work well in preschool. *However, this first AWTbook™ is not just for kids. I've written it for all deep thinkers, especially those with burning desires to know more about our place in Our Self-Assembling Universe.*

Preface

Part One – How to write an AWTbook™: It's safe to say that you are about to read your first AWTbook™. And I am about to tell you how you could easily be the first person to write another one. I will do that with the following, over-the-top, aspirational example.

Let's say you would like to create an outrageously wonderful life for yourself. Where one starts does not matter, but the way one starts could make all the difference.

1. Speak or manually write the following email-message to yourself with your smart phone, tablet or computer. "My name is ____. I was born ____, etc. etc." You will want such an introduction in the "About the Author" section of your book.
2. Create a title for your book. ***The Life of a Brain Surgeon*** is my example. You can either use this title or make up one of your own.
3. Move all such emails to yourself into an **archive** file on your device. The above message will be the first of many such messages inspired by videos or webpages that you will find interesting and of use in writing your book.
4. Having done the above, you can start the process by using your phone, tablet or computer. I have a voice-activate-able iPhone so all I need do is say ***"Hey Siri" and then say "YouTube - Brain Surgeon".*** One can also go directly to **YouTube.com**, click on YouTube's ***hourglass icon*** and type ***"Brain Surgeon",*** into YouTube's search field. Either method will instantly produce a selection of YouTube videos. The first one is usually the one you want but others that show up may also be of interest. You can also do something similar using other websites and completely different subject matters. I only chose Brain Surgeon as my idea of an over-the-top aspirational example.
5. Watch the video and then send a second email message to yourself. First, click on the feature called ***share*** that appears under the video, which will send the **video-link** to yourself having **shared** it with your own email address. This will leave a labeled and easily accessed **hyperlink** in your email that you can activate as many times

as you wish to get back to the video for future reference as you write. Some of these titles and links will also be the ones that you add to your AWTbook™. (This is precisely what I did throughout the book that you are about to read).

6. Also include in this email-to-yourself anything that you find especially interesting, such as new discoveries and things that you have learned from the video that you may want to explain later in words of your own choosing or that you may want to otherwise use in your writing.

7. Use key words from the videos to lead you to others. Some videos may even take you to subjects of great interest to you that are far afield from where you started and not at all what you were expecting. At any rate, at some point an idea should start to materialize to be the subject of your writing. This could also be the subject that you ultimately chose for a profession.

8. When I write I like to think of myself as an instructor who is trying to teach others what I have just recently learned. I did this often when I was teaching the new field of molecular genetics in 1970, and it is precisely the method I used when I wrote the current AWTbook™. I am not a nuclear scientist or quantum physicist, nor am I an astrophysicist, but I am someone who is keenly interested in the subject, and at age 83 I have clearly set myself on course to become one or the other. Due to my advanced age, success in that endeavor might seem to you to be rather dim, but I'm optimistic. No matter, it has been fun to write this book about these amazing subjects and I have learned a lot of cool stuff in so doing.

9. Give writing an AWTbook™ some serious consideration. I believe writing such a book may turn out to be one of the best ways to start realizing one's dreams.

10. The collection of emails that you have assembled, when reorganized in a sensible order, can serve in the same way index-cards do when one writes a standard book. Such organized files are used to create an outline that makes the flow of a book easy to write and, hopefully, easy to understand.

Part Two - Introduction: For billions of years Our Universe was filled with photons, and yet, for billions of years Our Universe was pitch-black-dark. There were only photons. *There was no light, at least not until the first, photoreceptive, sentient-beings showed up.*

Our Self-Assembling Universe*, OSAU-2, *C & I & L & E=mc²*, *is an AWTbook™ on nuclear physics, quantum mechanics and cosmology. It includes updates, clarifications, new information, and some of my own, recent, OMG revelations. The current edition also answers questions that might go something like this, "How the heck do we know THAT?!" But new to OSAU-2 are references to cutting edge **YouTube-type Videos**, **Audible-type Books** and **Podcasts**, which when backed by **Wikipedia-type Websites** can help clarify the subject, teach new things and inspire one to study.

***Part Three* – A new literary art form:** As just stated, OSAU-2 is introducing the AWTbook™, a first of a kind book that everyone **ought** to read because it's cool. Such a book takes full advantage of our new age of cellphones, tablets, computers and OLED smart-TVs backed by easily accessed **A**udio-based sites like **A**udible.com, **W**eb-based sites like **W**ikipedia.com, and **T**ube-based video sites like **YouT**ube.com.

For example, you will see in OSAU-2, which is now laced with references to YouTube titles, how the very best teachers accompanied by state of the art props, visual aids and film crews can quickly and effectively augment an AWTbook™'s thematic material. Difficult to grasp concepts can be realized in minutes by reinforcing subject matter with clear, entertaining, impactful, audio/visual demonstrations.

My thoughts on the AWTbook™ form-of-presentation have expanded as I write this. For example who might best write such a book? I have just written this one, the first AWTbook™, and I am no expert on the book's subject. But, what I **am** is a person who is keenly interested to learn the subject. Whether my lack of expertise is a good example, or not, might be up for grabs, but let's see what happens as the AWTbook™ concept evolves.

I can imagine many excellent writers finding themselves wanting to learn a new subject in depth, just as I have in writing this one. Take quantum field-theory for example. Some naïve-to-the-subject writer might want to approach QFT from a historical perspective, or they might just want to get proficient with the math behind it. Such a person, being an intelligent, perfectly suitable writer, starting from their naive level of understanding, would likely know best the questions to ask to get the answers most useful to anybody else who might be in the same boat.

For example, a writer might ask, "What is QED and who was first to come up with the idea?" They could start with YouTube, write about what they learn, add voice-activate-able **Titles** or photo-activate-able **QR Codes** as they go and thereby connect themselves and others immediately to the very best resources. Best of all their AWTbook™ could forever be a guide for themselves and others to a growing list of similarly linked resources. For example, I could see such a person writing a book with the following title, "An AWTbook™ on the Theory of Everything". They could even start from ***YouTube "We Might Have Just Found a Hidden Force of Nature",*** **(7:29).**

Part Four – Reality: When I began writing the first edition of *Our Self-Assembling Universe,* I had just experienced a "*Holy Yikes! Our Universe is Assembling Itself!"* revelation. It's obvious to me now. But I have to ask myself, *why indeed, was that concept a revelation?* And why did it take me so long to realize the truth about something which now seems so obvious? All I can surmise is that it must be due to the following. Every person I've heard comment about atoms eventually comes up with some version of the following stock phrase; *"Everything is made **of** atoms".* Even now, that is all I hear. Even the physicists I refer to herein say "We are made ***of*** atoms". But we are not just made ***of*** atoms are we? We are made ***by*** them!

The use of ***"of"*** rather than ***"by"*** may not throw anyone else off the truth, but I am pretty sure it had been instrumental in keeping me insulated and in the dark about the wonder of atoms assembling everything on their own.

You might argue, "Well, Frank, what about energy, thermodynamics, chaos theory, and the like? What about them or, even, what about God? Doesn't that definitely mean atoms are not doing everything on their own? And I say, "Well, just what do you think atoms are anyway if they themselves are not all of those things? And as for those with religious views, I doubt that any think God sits around all day telling each atom in our Universe just what to do, certainly not when a brilliant algorithm could do the job.

Maybe hearing the word ***"of"*** morphed into "***by"*** has given you pause or maybe it's even given you a *Holy Yikes* moment. Or maybe you've always just thought it to be obvious, or nothing that anybody should or could be thinking about. On the other hand, and this is what is of most interest to me, maybe there are some who think the concept of atomic self-assembly is total nonsense. I hope to dispel that position.

What all of you, dear readers, are about to witness is likely to be as startling as seeing a Lego Set assemble itself. You will all soon have the opportunity to witness and learn details of the way our Lego Set of a Universe has been, on a daily basis, pulling off just such a self-assembling magic trick. Everything, and I do mean everything, starting from day one, including the atoms, stars, planets, you, me and everything else is, right now, as we speak, assembling itself. And, it is doing so right before our very eyes!

You might even begin to realize, as I have, that we are part of the greatest magic show ever and it all seems to be based on some beautiful mathematical equations, one of which is $E=mc^2$. This equation, which leads up to the yet to be discovered theory of everything equation and its relationship to Consciousness (C), Information (I), and Light (L) is for me especially interesting. The significance of C, I, L, their connection to E, and the reason I placed my odd version of the $E=mc^2$ equation in the subtitle of this edition will be addressed along the way and at the conclusion of this Preface with reference to Max Tegmark and his book, ***Our Mathematical Universe***.

Part Five – How might one use this AWTbook™: Much of the information that showed up in the first edition of this book was acquired from ***wikipedia.org*** as well as a few other websites that I stumbled on in my early attempt to visualize how Our Universe came to be, and most importantly, how *we* came to be in this *particular* Universe. Since 2015, I have been through many books on astrophysics, quantum mechanics and the like. I've also frequently used and supported ***wikipedia. org*** because I found it to be a reliable, quick way to get answers for the initial questions I had when I started digging into cosmology, atomic physics and quantum mechanics. And I've used the expertise of Victor Troth and other physicist-types on ***quora.com*** to gain a little understanding of the difficult math behind particle physics. But then it occurred to me, for this volume I could do *something more*, and it just might be that that something more could turn out to be a game changer:"

I can now actively include titles to wonderful ***youtube.com*** videos that make it possible to relive, via filmed historical records, the key people and events that have been transforming our lives for the past 100 years in the rapidly developing fields of nano-technology. Many of the mind-boggling, quantum-level, scientific discoveries that have been made in *my* own lifetime have been filmed as they happened and are now available for viewing on ***YouTube***.

For a quick, 14-second example of the type of historical video to which I refer see:

YouTube *"(COLOR!) Albert Einstein in His Office at Princeton University", DRsewage (0:14).*

And then maybe much later after you know a thing or two see:

YouTube *"The Complete FUN to IMAGINE with Richard Feynman", Christopher Sykes (1:07:00).*

Regarding these videos, I've been particularly impressed by the excellent teachers and state-of-the-art pedagogical tools that one can find in many of them. Even with all of my listening and reading, it's only through the explanations delivered by these experts that I've been able to fully appreciate and actually begin to understand the complex, nano-technological discoveries that now shape our lives. Also, because I've seen how the videos, which I've referenced herein, already do a good job delivering much of the knowledge and clarity that I was hoping to deliver in this volume, I now think OSAU-2's most important value, other than its introduction to the AWTbook™, resides in its ***simple, classroom/at home, tactile, teaching-tool approach***.

But, because this book is an AWTbook™, it is a special tactile-tool backed by web pages that contain accurate, detailed and necessary explanations as well as YouTube videos with great teachers armed with stunning audio/visual aids. I've found these web pages and videos bring excitement and deep understanding to the fascinating nanoscale world of molecular machines.

Moreover, once students view even just a little of the nanoscale, world of atoms, I've found many of these students to be highly motivated to learn more by the visceral experience they get when they assemble models of atomic nuclei on their own. The videos that I have referenced and interspersed throughout this book will show you what I mean. These videos really nail it.

Although you might find the content of some of the videos to be beyond your immediate understanding, I think by the time you have consumed OSAU-2, it will dawn on you that you understand that content and, with an excitement akin to my own, finally "get" what Democritus, Aristotle,

Copernicus, Galileo, Newton, Faraday, Planck, Einstein, Heisenberg, Bohr, Schrodinger, Feynman et al were trying to "get" for themselves and have been trying for years to "get" us all to understand.

YouTube "Intro to History of Science: Crash Course History of Science #1", CrashCourse (12:20).

Oops, I left out Archimedes. To make it up to him, have a look at the following presentation.

YouTube "How taking a bath led to Archimedes' Principle – Mark Salata", Ted ED (3:01).

And if you want to know where Democritus has taken us, have a look at **Wikipedia, *"Quantum Computing Since Democritus*" Scott Aaronson (398 pages).**

Regarding excellent teachers, **YouTube** has many. Physicists **Jim Al-Khalili** is but one example to which OSAU-2 gives reference. You will find him to be a great writer and an amazing storyteller. Teamed up with excellent producers, and state-of-the-art videography, his videos bring details into breathtaking view of the creation of Our Universe by the self-assembling, nanoscale robots we call atoms. Eventually, I think you may want to watch all of his many videos.

Given the new audio/visual AWTBook™ dimension added to **OSAU-2**, there are now at least four ways readers can use this book. With computer, laptop, or as I like to do it. With voice-controlled mobile phone and headset in hand, one can either *1. Quickly read through this book cover to cover and then check out the referenced videos, 2. Do it the other way around, 3. Do it stepwise by viewing the videos as one goes, or 4. Do it, as I suspect most will do it, by checking out the short videos and selected portions of the long videos as one goes.*

No matter the way, having done the reading and viewing, one should be able to fully appreciate OSAU-2's **tactile approach** to understanding Our Self-Assembling Universe as one assembles atomic nuclear models of it with inexpensive ping-pong balls and fabric tape. Not only that, I think anybody, having thoroughly devoured all of the content of this AWTBook™, will be well on their way to a career in astrophysics, quantum mechanics or molecular genetics. How so? The content, herein, is vast with its many outstanding YouTube videos and Wikipedia explanations. But one video will no doubt be inspiration and motivation to see the next where one can see complex concepts and scientific tools expertly communicated in a matter of minutes in great precision and breathtaking detail.*

*****I'm no tech expert (understatement) so I am sure there is a better way to do this, but I have found the following to be a good way to store links to written material. With cellphone in hand I hold my finger to the proffered link until it reads "Copy" which I then select by touching "Copy" with my index finger. Then I "Paste" the copied link to an email that I send to myself. I swear, the phone remembers that I held my finger to a link that I wanted copied! And my phone will remember it for a while until I press my index finger to the blank body of an email to myself until my phone reads "Select", "Select All" or "Paste". It's like magic, when I touch "Paste", my phone copies the link to my email. I can collect multiple links this way that I individually label and send to myself to store in "Archive", "Saved messages" or "Drafts". Alternatively, for YouTube and some other sites storing links is much easier. I can just use the site's ***share*** command to send the link to myself. This technology is rapidly evolving so I expect all of it will get to be a lot more consistent and easier in the future. Until then, maybe my struggle with this will be helpful to some.*

Part Six – Back to Reality: "Okay", as the old radio show used to say, it's time to "Stop the Music!" Right this minute I want all of my readers and listeners to stop what they are doing to see what I mean. I want everybody to get a visual, mind-blowing sense of the *real* magic that is right now taking place in Our Universe, and to know why I think it is important to use the term ***made*** "***BY"*** instead of ***made*** "***OF"***.

I first wrote OSAU in 2015. It is now 2022 and some unbelievable discoveries and some equally unbelievable technological advances have taken place. It turns out that seven years is a long time when it comes to our modern technological advances. Many of these advances, which made possible the following impossible-to-believe video, took place in that time. So, please, stop what you are doing and take a long look at the **short** video by **Derek Muller,** which is just one of the many great videos that Derek offers on his *Veritasium Channel.*

YouTube *"Your Body's Molecular Machines",* (1:20).

As I am reading my AWTbook™ with my iPhone in hand I can simply say to it, "Hey, Siri!" - - "Your Body's Molecular Machines", and up pops the link! I don't even have to activate my phone, my voice wakes it up. There is usually not even any need to refer to YouTube. YouTube almost always shows up as a choice this way but you will also get Wikipedia, etc. choices. So, to avoid that option just say, "Hey, Siri, YouTube, Your Body's Molecular Machines". Now you will get a selection of thumbnail videos, of which the first one is usually the one you want.

Now that you have seen Derek's video, are you not now gasping? Do you not now want to see, read, listen and learn more having watched cell-division accomplished by amazing nanoscale robots assembled by the very same atoms of which *they* and *we* consist? Do you not now clearly understand that these nanoscale machines are at work in you and are, right now, *on their own,* accomplishing the seemingly impossible? Are you not, right now, chomping at the bit to write your own AWTbook™?

These representatives of molecules in action have been accurately created, realized and animated based on years of experiments and observations. *AI*-aided electron microscopy, atomic-force microscopy, and state-of-the-art, X-ray-diffraction crystallography, all backed by over 100-years of biochemistry, genetics, quantum physics and 40-years of molecular genetics, have come together to create the animated images that now let us see, for the first time, nanoscale-robots doing their thing.

The above short video will lead you to other videos like it, some of which are much longer and far more detailed. Many of these remarkable videos show molecular machines of various kinds operating in real time. See, for example:

YouTube *"mRNA Translation (Advanced)", DNA Learning* (3:03).

YouTube *"The Central Dogma of Molecular Biology", Khan Academy,* (4:22).

YouTube *"Electron Transport Chain", HarvardX* (7:45).

YouTube *"ATP Synthase in Actionz", HarvardX (4:59).*

YouTube *"Mitochondria: the Cell's Powerhouse", HarvardX (5:17).*

YouTube "Bacterial Flagellum –A Sheer Wonder of Intelligent Design- video", Philip C (7:37).

Phenomenal, are they not? Are you not now blown away by the way us eight-billion or so humans so miraculously showed up? I am, especially when I take the time to think about myself at this time and at this place riding on the surface of a large rotating ball that circles a star in a whirling galaxy of 300-billion or so other such stars that race away in Our Expanding Universe along with trillions of other galaxies at a rate that some estimate to be about 1,300,000 mph.

We are, indeed, astronauts kidnapped on a spaceship that is now traveling through space at an incomprehensible speed in an incomprehensibly enormous, expanding Universe! And, while we are at it, we might as well completely blow our minds with the wild things that go on in Our Universe as we travel, such as the super black-holes that galaxies have at their centers.

I'm probably not correct, but I like to think of the super black-hole that exists at the center of our galaxy as a mega drain that drives our swirl much like water draining out of a bathtub. Everything spins about mass in Our Universe so the tub-drain image doesn't appear to be a popular explanation for our galaxy's spin. But what's our black-hole up to if it doesn't have anything to do with the apparent fact that it seems to be eating stars like water flowing out of a bathtub? And, let us not forget the exploding supernovae, colliding neutron stars, pulsars, quasars and the realization that all of the violence these things create is responsible for our kidnapped existence, as well as our silver, gold, and platinum-diamond jewelry. Also, let us not forget to thank our kidnappers. I speak of the atoms and their quantum-world transformations, the molecules, that have been taking us on this trip, *whether we like it or not I might add*!

YouTube *"Why is the Universe Perfect?"*, (34:30).

The latter stunning video summarizes pretty well my idea about **C & I & L & E=mc²** and the incredible journey on which such equations have conspired to take us. But there is more. The following video is so mind blowing I will not be surprised if it doesn't blast your present mind out of its housing!

YouTube *"Time-lapse of the Entire Universe"*, (10:49).

With OSAU-2, I hope to bring clarity to the spaceship trip that we are on, at least as I see it. Although I'm not going to spend much time on the controversial C-part of the equation, I think it might just provide a way for everybody, religious or otherwise, to eventually understand Consciousness in a unified way.

The first law of thermodynamics states that Energy is neither created nor destroyed. My connecting of **C** to **$E=mc^2$ in C & I & L & $E=mc^2$** simply hypothesizes that in some *chicken versus the egg style* the whole equation*, if appropriate units can ever be created and rationalized in some way*, is conserved, including **C**. That is, and this might hurt the brains of a few people I know. **C** is neither created nor destroyed in **Our Universe**. Right or wrong, I think the following three things are self-evident and provide helpful clues as to why I think that C, at least in Our Universe, is as conserved as E;

1. We humans are not alone in Our Universe.

2. Many of us on this comparatively young planet have come to be completely absorbed by our love for video games and virtual reality.

3. Humanity is on the cusp of events that will leave some of us, *currently* alive on this planet, *living for a much longer time than most of us have imagined*. Refer to Ray Kurzweil's video.

YouTube "*Kurzweil Discusses Living Forever", James Bedsole (2:50).*

Put all of the above together and think about it. What would *you* do if you found *yourself* living for a very long time, maybe even, for as long as you like? In the relatively near future, quite likely in less than 10 years, a good guess is that experiments now being done in animal models, *which clearly suggest that aging is a curable disease*, will be confirmed in humans (Refer to **David Sinclair's** work below).

Also, within the next 10 years it is almost certain, according to Kurzweil, *AI* supercomputing singularity will have arrived to send us well on our way to a Star Trek reality. Can anyone say "full emersion simulation"? Just saying. This simulation idea is key to understanding my view on C=I.

And to understand how my idea fits into **string theory** and the **multiverse**, I recommend the in-depth series of **YouTube lectures** by **Leonard Suskind**. The following video is a short discussion of the way **string theory** with its infinite **landscape** of **multiverse** possibilities can explain **Consciousness**. Could be, but I am not relying on that theory. I am relying on a proof that requires **I (I**nformation/**I**ntelligence**)** come before **C** in **I=C** where, when it comes to **C,** we are but infinitely tiny examples. However, I'm not saying that we are not important. Why not? Because we are the most easily identifiable proof that **C** with at least a modicum of Intelligence, exists in Our Universe. In my scenario the directionality of I=C comes directly from the seemingly certain fact that we humans will create an artificial, silicon-based **"I"** that leads to a life-form **C** as real as our own **C**. We may even do that "very soon" aided by AI Singularity *(or, if not us, others like us doing so elsewhere in Our Universe)*.

Youtube "Life in the Universe. A Journey to Outer Space", Kosmo (1:24;32).

Needless to say, I am overwhelmed by the thought that we have, in effect, been kidnapped here at this time and place, and at the whim of some very imaginative, extraordinarily strange, supposedly inanimate nanobits of matter that are armed with the rules and skills of quantum mechanics, which make them, don't you now think, way more animate, wavelike and mysterious than is humanly possible to understand?

This gets me to the second and most recent revelation, and the other reason I am so excited to be able to resubmit *Our Self Assembling Universe* with this Preface. I think, because of all of the quantum magic and mystery, few of us, and that might actually mean none of us, have any idea what we are talking about. For example, I just said atoms are "things" like I know what I'm talking about. I really don't, and I don't think anybody else does either. Not only that, we all talk endlessly about other things that we think exist and with which we think we have knowledge and experience. But do we even know what this thing is that we call "ourselves"? Do we even have a clue about that? Maybe, but I wouldn't be surprised if we were wrong.

So, all that the unknown stuff means to me is the door is wide open. I invite everybody reading OSAU-2 to walk through that open door, especially all of my young readers. Go out, take the chance of making a fool of yourself, just as I am now doing, and discover something. It's fun! There is a lot to uncover and create! We are going to Mars and I look forward to seeing you there.

YouTube "Elon Musk Will Reach Mars with This Amazing Rocket Engine", Daily Aviation (11:37).

Part Seven – A Cliffhanger with a Question: *Before going on, just for fun, here is a cliffhanger. Many know what the speed of light is when light finds itself in a vacuum, but how many know what light's speed is when it is not in a vacuum? For example, at what speed does light travel when it is taking a trip through water? The answer to this question might come as a surprise to some. Look for the answer below in my, now, still feeble, but less inept, attempt to understand the weird properties of light.*

Hint, the answer to this question is proof positive to what I think should be obvious. ***C=L.*** *But please note. I have no idea what the units should be for such an equation,*

Part Eight – Listening vs Reading: Since 2015, I've been listening to and reading books trying to get to a better understanding of my "OMG!" revelations of *"Our Universe is Assembling Itself!"* and, *"Yikes!, nanoscale robots are, at this very moment, reassembling me!? And their actions are the only thing between this thing I call myself and oblivion!!?? You have GOT to be kidding me!!!"*

My list of readings began in 2002 with the only science-history book I've ever read from cover to cover, twice. Why? I had the time. I had just retired from the company I co-founded in 1982 and had plenty of time to read something more than endless scientific papers. I actually had time to read a real book!

But, and this is what is really important, I wasn't done with that book. It also became the first book of many that I have since listened to from cover to cover, twice. That first book, "***The Making of the Atomic Bomb"***, may I say, is wonderful. I recommend buying the book written by **Richard Rhodes**, reading it from cover to cover, keeping it as a reference, and then getting the audible version and listening to it several times. I found it to be even better to listen to the book, as it is read by **Holter Graham,** one of the great narrators in the world of **Audible Book** story-tellers. All other books in my list of "readings", which I have employed in writing this preface, I have not read. But I have listened to every word, and may I add, with proper emphasis and inflections delivered by great narrators. However, for almost all of them, I have also purchased the hard copy, which I keep for reference, diagrams and pictures.

I'm adding one more book here that you might want to read or give a listen someday. I just picked it up while writing this Preface and found it to contain a wonderful introduction to quantum mechanics. The book, "***Einstein, His Life and Universe"***, written by **Walter Isaacson,** is narrated by **Edward Herrmann.** The first chapters are very humanizing as they share Einstein's youth, loves and failed relationships. They make Einstein into an approachable, real, fallible and, for me, likeable person. Starting with chapter five, Isaacson does a fantastic job of explaining Einstein's life and the details of his science during what are aptly referred to as his **"miracle years".** Another cool thing Isaacson does in describing these years is dispel the misleading picture of Einstein surrounded by blackboards filled with equations. Blackboards did end up filled with unbelievably complicated math but that mostly happened much later when he was nearly driven crazy working out the math for his **Theory of General Relativity** and even later with his frustrating failed attempt to work out the math to fit his **Unified-Field Theory of Everything.**

Listen to chapter **five** and **six** and you will understand what I mean about Einstein's blackboard-free-of heavy-math process. Even though math is a big help when it comes to *deductive* as well as *inductive* discovery, and even though math may be THE key to understanding it all, I think it is wonderful to find out that in order to be an "Einstein" it would not have taken heavy math to *deduce* many of the profound truths of Our Universe. Nor to be an Einstein does one need empirical, lab-based experiments, even though there is no doubt that controlled experiments can be a big help when it comes to inspiring *inductive* reasoning and *deductive* breakthroughs. And controlled experiments are absolutely necessary when it comes to establishing proof. Nonetheless, there is one thing that doesn't require any of that and it is for certain *the* thing that one must do to be an "Einstein".

What is that thing? It is, simply *deductive,* no-benchtop-work-required, ***thought-experiments***. I will share with you examples of Einstein's **Thought Experiments** on **Special Relativity** and **General Relativity** below. I will also share with you an example of an **Empirical Benchtop-Type Experiment** which, if you had devised 3000-years ago, *could* have led you to a **thought experiment** that *would* have made you amongst the most famous scientists who ever lived.

Isaacson's book is chronologically interspersed with chapters dedicated to Einstein's science. In my opinion these chapters contain some of the best historically-based descriptions of Special Relativity, General Relativity, Quantum Mechanics and Unified Field Theory. All of these chapters can be visually augmented *AWTbook™-style* by using key-words from those chapters that will bring you access to outstanding video presentations. And if you are turned off by the love-crazed antics of a surprisingly human Einstein, you can simply skip those chapters. I didn't. I thought they were interesting interludes that helped bring my picture of an unapproachable great man down to Earth. The science is to be found in chapters 5, 6. 7, 9, 11, 20, 21. The other chapters deal with coming of age, politics, social issues, strange things and some earth shaking events.

Before I get off the topic of listening versus reading, there is a huge health benefit to listening, but only if you promise yourself that you won't get to hear what happens next unless you are moving. As I write this preface at age 83, I am in fantastic shape. I run, jog, and walk five to eight miles most days, and have done so for the past 15 years just by using a single question for motivation, "What happens next?" For example, this morning it's 7 am and I am excited! Why? Because, I just watched one of **Jim Al-Khalili's Spark** videos,

YouTube "*The Story of Energy with Professor Jim Al-Khalili / Order and Disorder", Spark (59:00).*

That video just now got me excited to listen once again to his very cool AudioBook, "***Quantum: A guide for the Perplexed"*** read by a great narrator, **Hugh Kermode**. I can't wait to get outside, start listening and start walking, jogging and running. This audiobook is great and, even though I've heard it before, I will find myself excited when I wake up tomorrow morning to hear what happens next. This life-affirming motivation for exercise is amazing. And it's *so* simple.

One more thing while I'm at it. Maybe you wouldn't want to live forever, as alluded to above, but I'm guessing you might like to live a long life in good health. If so, you should listen to or read "***Lifespan"*** by **David Sinclair**. The author, himself, is the very excellent narrator. I have the hard copy but I've already listened to the audible version twice. It's that good. The unbelievably important molecular machines that you will learn about are the enzymes called ***sirtuins.*** The molecules with which sirtuins interact are the ***histone*** proteins referred to in the *molecular machine* video you just watched. Histones are the proteins that DNA wraps itself about to form ***nucleosomes.*** Nucleosomes are the globular bodies that comprise the **chromatids** that comprise the ***chromosomes*** that you saw in the video. To see DNA in action one more time as it ***wraps itself*** about histones, please watch again **"*Your Body's Molecular Machines*" (6:21)** and then follow that up with these short videos:

YouTube "*How DNA is Packaged (Advanced)", DNA Learning Center (1:43).*

YouTube "*Epigenetics Overview", Cell signaling Technology, Inc. (2:15).*

YouTube "*Life in Miniature: Our Secret War Against Cancer", Sarah Gillespie (3:11).*

These videos should be enough to convince anyone that we are constantly being re-self-assembled by atoms. And the above video will be of help when it comes to understanding David Sinclair's descriptions of sirtuin enzymes and their actions. I think knowing about one's sirtuins is not only important, this knowledge may lead others to learn more about **epigenetics** to discover more ways in which we can help the enzymes that run the magic show of reassembly that goes on within us each and every day. For example, one can learn all about **sirtuins** by using this AWTbook™ and its *Wikipedia-type, YouTube*-type approach. YouTube constantly updates certain videos and those by **David Sinclair** are especially noteworthy. The following is one of David's most recent and is an example of benchtop, inductive research leading to a potentially life-changing theory. He also just started a podcast with his co-author that can keep one up to date on the latest breakthroughs in aging research.

YouTube "*David Sinclair New Aging Reversal Approaches 2021 Update", Dr David Sinclair* (20:20).

Youtube *"Double Lifespan-David Sinclair", Serious Science (12:15).*

The disease of aging is a field of research that is hot and rapidly moving. By the time you read OSAU-2 David and others will have taken us a lot further on their missions to expand human lifespan. So, look for **the most recent** publication dates and refer to his YouTube podcast.

Part Nine – How it all began in 1900; Okay, now I hope I have your interest. How *do* we know that the solid stuff in Our Universe started from something we call atoms about 13.8 billion years ago? And how could we possibly know anything at all about the beginning when it happened so long ago? A lot has been learned over the past 2000-plus years of us humans searching for answers to such deep questions. And it's the early questions and learnings that led to the above astounding revelation. Over 2000 years ago **Democritus** (460-370 BC) and others were already having deep thoughts. One of which was, *of what is all this solid stuff made*? Democritus even gets credit for the name **atomons** that was given to his idea at the time, i.e., his idea for an invisible, indivisible tiny-stuff-composition for the solid stuff. The following is the first of a very useful series of videos. This first one starts with Democritus.

YouTube "*What is an Atom and How Do We Know?", Stated Clearly (12:14).*

As the above video shows, Democritus had no real evidence for his idea, but neither did anybody else until "very" recently. Moreover, we didn't stand a chance to know if Democritus or the two famous atomists who came after him, **Isaac Newton** (1643-1727) and **Ludwig Boltzmann** (1844-1906), might be right or begin to understand what *might* actually be going on in our Universe, at least not until **Max Planck** (1858-1947) came along. Planck devised a math formula in 1900 that was so revolutionary he, himself, did not believe it! I think it incredible, but even as late as 1900, **Max Planck, and most others alive then, still did not *believe* that atoms existed**. So it is no wonder that a little math formula might be hard to swallow when it insists that light's a particle, especially when everybody *knows for sure* that light's a wave.

However, whether it was believed or not, Planck's formula worked. It turned out to be the only mathematical statement that *would ever* be able to rationalize how light changes color when black metal objects are heated, or explain the nature of radiation issuing forth from a lightbulb. And if Planck hadn't been given the task of figuring that out for a lightbulb company, we might still be in the dark.

The following video is a good history of the science leading up to **Planck**, and the development of the wave equations of **Ludwig Boltzmann** and **James Clerk Maxwell** (1831-1879), which are used to this day in the sciences of thermodynamics, field-theory and quantum mechanics.

The narrator does a good job in helping us understand what Planck had to deal with when he came up with his revolutionary discovery.

***Youtube "Boltzmann's Entropy Equation: A History from Clausius to Planck*", Kathy Loves Physics & History (24:34).**

So it was that in 1900 everybody "in the know" at the time, including Planck, *knew for certain* that light was a ***wave***. However, even though that is what he thought he knew, his formula insisted otherwise. There was no other way to understand the outcome of his discovery. His formula insisted that light must have properties akin to a ***particle*** **-** at least somehow, sometime or someway. But instead of going along with what he found, he must have thought something like, "Oh well, it's just a funky bit of math that helps me understand a few things useful for that lightbulb company". Or, and maybe this wrong thought is closer to the truth of his thinking, "The effect has nothing to do with light itself. It has to do with the effect that light has on the things with which it interacts." At least, from what I've read, both of these thoughts could well have been his reasoning because he fought so hard against the idea that light was in any way a particle. He and most everybody else at the time *were absolutely sure* that light was a wave, and that Our Universe was filled with some kind of **"*ether*"** that was needed to support light-waves. It is no wonder then, with all that misconception going on, that he, too, was stuck on the wave idea. However, it matters not what everybody thought. Planck's formula and the change in thinking that it suggested led to the development *during his lifetime* of a new branch of science called ***quantum mechanics***, the science that attempts to explain how everything "looks" and "acts" at the quantum nanoscale.

If you have gotten the idea that by 1900 EVERYBODY in the developed world had all of sudden become very interested in light, you'd be right. *(The following video by Al-Khalili is special because it answers questions about how we know stuff about stars and the age of Our Universe. I've posted it here, and later in this Preface where it is likely to have greater impact.)*

YouTube "*Light and Dark, Both Episodes / Jim Al-Khalili", Abubakar Hamid (59:34).*

Max Planck's revolutionary formula ***E=hυ***, states Energy, ***E***, is *proportional* to the frequency, υ,of light. And, most importantly, his formula fantastically states that υ, the frequency of light, is *equal* to

E when it is multiplied by a *vanishingly small* constant, ***h***, now known to have the coveted standing of ***universal constant***. **Planck's too tiny to imagine *h*, 6.62607015x10^{-27} cm^{2} grams/second** is used in all thing's quantum mechanical.

Since Planck's formula insists that light must in some way act as a wave-like *particle*, the formula can be rewritten with an ***"n"*** as in ***E=nhυ*** to represent the number of particles involved in any particular reaction involving light. This is a very simple yet profound equation. And, as Planck himself would say, "it is-beautiful". It is, isn't it? But to understand the formula so we can better understand how our Universe works and how quantum mechanics came to be, we need to understand υ, the frequency of light. And after that we need to understand ***h***, Planck's deceptively ordinary looking, Universe-Assembling-Itself-Constant.

At the start, but only at the start, I like to think about light waves in terms of ocean waves. Such waves, as is true for all waves, have high points, the crests, and low points, the troughs. The difference in height between the troughs and the crests is the *amplitude* of the wave. One way to measure the frequency of in-coming ocean waves is to head out to the large swells. Once there, in order to gauge their frequency, we can time the crest of each swell - just before those swells collapse and the waves smash into our frames.

But now we are in for it! Our goal has changed to a frantic attempt to avoid drowning. The ocean is rough and the waves are hitting us at a frequency of one a minute. Well, at least we got what we came for. That's all there is to it! A wave a minute is the frequency that day of that set and at that time for that heavy surf. Measuring light waves is no different when it comes to calculating frequency.

But light waves are VERY different and they are weird, as you will learn, and they are of course not very big. They are nanoscale-tiny-things with quantum-particle properties. Also, just to be different, scientists have given electromagnetic waves that arrive every second a specific **unit of measure**, which they, and now we, will call the ***hertz***. That's not a rental car, it's a unit of measure that is, simply, equal to a wave of light or electricity coming in at some multiple every second or, as those in the know, and we will now also say, in ***cycles per second.*** BTW, the *hertz* is named in honor of **Heinrich Rudolf Hertz (1857-1894),** the first person to provide conclusive proof of electromagnetic waves.

So, even today, light is most easily thought of as a wave and atoms are most easily thought of as particles, i.e., solid objects, but light isn't exactly a wave, and atoms are certainly ***not*** solid objects. In reality light comes in particle-like packets of energy called **quanta**, AND SO DO ATOMS! That is really hard to think about but maybe this will help. I like to think of quanta as **nanoscale versions** of the **forcefields** I used to read about in science fiction. That makes sense to me because if atoms are like ***Sci-Fi forcefields***, energy quanta only create the illusion of something that acts like it is solid!

I am sitting here, right now, writing this Preface. I'm surrounded by a world that appears to be quite solid. ***It is not.*** It is all an illusion created by atoms that are nothing more than nanoscale forcefields or as *those in the know* call them, atomic energy quanta. Thus, we begin again to try to understand all of this as we go back to Max Planck. But before we do that, maybe I have left you with a misconception. In-coming waves of light are *totally* nothing like ocean waves.

A thought experiment by H. Frank Gaertner (Spoiler Alert, it's only my effort to help me get closer to an understanding of quantum weirdness. My "experiment" does not really explain anything. However, if it helps you understand the weirdness in a likewise manner, mission accomplished).

A beam of light used to be thought of as something akin to ocean waves and other macroscopic waveforms, but since light travels in particle-like-**energy packets** called **quanta or photons,** and travels in a vacuum at 186,000 miles per second in straight lines, barring quantum gravity effects, I think a beam of light might be more analogous to shot-pellets that blast forth from a shotgun, but with some odd twists. For one of these twists you must think about shotgun cartridges loaded with a new type of spinning, rubber-shot-pellets armed with grinding barbs where the pellets spin in a wave-like equation-form that varies in intensity with a strict dependency on the pellets rotational frequency.

The gauge of the rubber-shot does not change, but if you want the greatest devastation you buy the new cartridges with barbed shot having the greatest rotational frequency. Got it? If so, what follows is the thought experiment, which was mine and is now yours.

You shine a beam of light from an ordinary flashlight on a wall. You draw a circle around the circumference of the beam as you see it on the wall.

Then you take your shotgun and fire at that circle. Having backed away from the wall to the right distance, the shot pattern closely matches and pretty much fills the beam's circle. So far so good. Neglecting speed, since pellets from a shotgun move much slower than light and initially accelerate and then decelerate while light's speed is a constant 186,000 miles per second (another twist), light and the shotgun blast compare pretty well as they fill up the same area of the wall. Moreover, if the comparison is going to continue its approximation, the shot has to be special as described above. Barbed shot can be made to spin at any frequency you desire and all of your desired frequencies are available for purchase. Thus, the damage done to the wall will depend on the frequency of rotation and grinding energy delivered by the spinning barbed shot in the specific cartridge that you fire each time at the wall. Remember, no such shot currently exists. You just purchased it and have no idea how to make it barbed or make it spin, but it's not too hard to imagine how it works. PS, the shot itself is a rubber pellet that has no damaging effect.

Your analogy is not exact but, at least, it gives you the feeling that you might have a partial understanding as to why the frequency of light is important when it comes to the difference between the destructive powers of ray-guns versus flashlights. But you must not forget, ***A=Amplitude****. The more shot you have to fire in your cartridge the better for damaging the wall with spinning, barbed shot. Likewise, you realize that the more photons fired, the better it is to damage a wall with a ray gun.*

But hold on, you think, I still don't have a great analogy with the shotgun when it comes to understanding the all-important ***superposition, entanglement*** *and* ***coherence*** *of* ***photons*** *that is needed for the* ***Amplified Stimulated Emission*** *that exists for light emitted from* ***lasers,*** *and you realize that's another twist that makes light different from your shotgun blast since* ***coherence****, which permits the* ***amplified stimulated emission*** *damage that can be caused by a light beam from a laser gun is not possible when it comes to any shotgun blast. You give up at this point and decide that maybe more can be revealed when it comes to the subjects of* ***superposition*** *and* ***coherence of quantum particles*** *with another thought experiment.*

But even then you realize you have another twist coming. ***Light is known to be an electromagnetic wave, where the straight line***

course of light is moved off its course into waveform by a vertically acting magnetic force and vice versa. In other words the electric wave and the magnetic wave are in phase. It's the oscillating interactions of the electric and the magnetic forces that create the familiar sine wave representations of electromagnetic light beams. And if you happened to view the video I posted at the beginning of this preface, you will know of that to which I refer.

So you might ask, how does a rotating barbed rubber pellet fit into that model of light? ***?*** *It doesn't of course.* *And I'm not sure my analogy helps. If the* ***pilot wave*** *idea of* ***De Broglie*** *and the* ***silicon droplet*** *experiment have anything to do with quantum reality, we have a horse of a different color.* The first short video shows the silicon droplet experiment**.**

YouTube "*Is This What Quantum Mechanics Looks Like?"* Derek Muller (7.41 min) .

Others like this, if you want more, are:

YouTube "*Through the Wormhole- Wave/Particle – Silicon Droplet*" bajarwas (4.1 min).

YouTube "***Do We Have to Accept Quantum Weirdness? De Broglie Bohm Pilot Wave Theory", Looking Glass Universe (12 min).***

See also, Julian Schwinger's version of QFT which, I just discovered, explains this paradox as units of light existing as a *quanta spread out in space* until they collapse into photons when they are detected.

I know, I get sidetracked when I think an important idea might stir things up. So I thought it important to bring the thrill of experimentation, which I myself have experienced, by suggesting some quasi-thought experiments to the benchtop uninitiated where such an experiment might apply to discovering constants like the one Planck's laborious mathematical exercise pulled out of the hat to arrive at ***h,*** a constant so important that it unwittingly allowed Planck to introduce Quantum Mechanics to the world in 1900.

I think it is cool when a great scientist shows that they, too, can miss the big picture when it comes to discovery, just as I did so many times in my own research career. However, I must say, I never discovered anything so colossal only to miss the big picture, as Planck did, by thinking my finding was nothing but a convenient math trick. Fortunately, Planck reluctantly, *sort of*, changed his mind when Einstein came up with his famous, Nobel Prize winning, photoelectric-effect experiment.

Einstein, the theorist and expert-thought-experimentalist that he was, never ran the actual benchtop experiment but he did think of it, describe it in detail and did show in a research paper how his photoelectric effect experiment was to be done. Many others since have done Einstein's experiment to conclusively prove light's particle-like behavior. And, as luck would have it, it was Planck, when he *began* to see the "light", who helped the, then, struggling, lesser-known Einstein gain his future, great, scientific respect and worldwide notoriety.

And now that we, too, know that light behaves like a particle let's really take some time to *try to* understand ***h***, a **Universal Constant** that's now known to be THE unit of energy which is involved in ALL of the single, non-continuous jumps or quantized changes that take place in the *angular momentum* of Our Universe. And, it seems, if I have it right, that means *all of the energy in Our Universe can be understood in terms of angular momentum*. And, of course, that means we are also saying, **it's *all*** about circles, oops, I mean waves. If you ever wondered what trigonometry had to do with it, now you know. To wit, what are photons, electrons, protons, neutrons if they are not all related to spherical objects or forces? What generates a sine wave? Could it be a circle? Just saying.

YouTube *"Sine-Curve and the Unit Circle", Arkady Etkin (0:56).*

***YouTube "Sine and Cosine from Rotating Vector",* Khan Academy (3:56).**

And if you ever want to get the total energy of the quantum particles in Our Universe, you now know that you will need to multiply ***n***, the number of those particles, by something. That something is the ***h***υ of Planck's formula (*after **h** has been modified for quantum spin, and the possibility that the number of particles might need to include virtual particles*).

YouTube *"Are Virtual Particles A New Layer of Reality?", PBS space Time (17:14).*

In an easier to understand system, for example, one just multiplies ***h***υ by the number of photons in a beam of light or the number of electrons in a charge to get the answer to the question of "what is the energy in a beam of light or an electric charge?"

Back to your thought experiment. I think a Planck-like formula could be developed for the blast from your special shotgun pellets. The increase in rotation that one gets with the purchase of more lethal versions of your imaginary rotating-rubber-shot cartridges is like the increase in wave frequency when you use shorter wavelengths. Also the number of times you fire your shotgun is **n**. And the number of shot in the cartridge is **A,** the **A**mplitude, limited here by cartridge size, which is similar to increasing the brilliance of light by increasing the number of photons in the flashlight beam limited by its brightness control. The increasing frequency of rotation and destructive level of the shot purchased compares to the increasing frequency obtained when one uses shorter-wave-length-light to deliver higher, more damaging frequencies. The potential for damage can increase with the speed of shot rotation, just as the potential for damage can increase with shorter wavelengths of light. One way to see the damaging effect of shorter wavelengths of light is to use a laser. See how we just moved from particle behavior to wave behavior?

I think it might be interesting to bring up something that is probably obvious to most but deserves special attention because I think it represents a very special thing about these important constants and their equations. The important constants can find themselves in simple yet very important relationships. For example, two of the most important ***Universal Constants*** derived in recent years can be found in Planck's and Einstein's most famous equations.

These two equations and many others of nearly equal, or maybe even greater importance in everyday use are nothing more than simple proportions. And proportions can also always be expressed as ratios between variables. For example, $E=h\upsilon$ or $h=E/\upsilon$, $E = mc^2$ or $c = (E/m)^{1/2}$. These simple relationships, with their **constants** of proportionality expressed as a ratios between variables, sometimes are of the Universal, unchangeable kind, such as is true for ***h*** and ***c***, but many more times such constants just involve useful relationships. For example, the equation for the circumference of a circle contains pi, an irrational number, but still a constant, $C=\pi D$; Newton's equation of force; $F=ma$, where one can hold either ***m*** or ***a*** as the constant to get ***F*** is very useful. Whereas, Newton's **G** in $F=GMm/r^2$ is universal.

It's ***the*** Universal Gravitational Constant. Then there's the ideal gas law, ***PV=nRT*** (see **Wikipedia**, ***"Ideal Gas Law*"**), where ***R***, or at least a modified version of it, is the Universal Ideal Gas Law Constant. Now comes something about these highfalutin constants that I find especially interesting.

All the above equations produce straight-lines when one graphs them on Cartesian coordinates. That is, the above equations are all in the form ***y=mx***, where ***m*** is the ***slope*** of the straight line that's equal to the amazing ground-breaking constant that's connected to each of the above equations. That's it, except to note that the straight line does not always have to go through zero. In that case the equation will have a Descartes Coordinate intercept that's other than zero, which simply alters the equation to $y = mx + b$.

Therefore, if you are a benchtop, inductive-type experimenter with an unknown and your empirical experimental data can be represented as a series of points that match a straight line on a graph, you just might be onto something. Moreover, the slope of your graph might turn out to be heading you in the direction of a ***constant*** of some **great importance**. Interesting, isn't it?

YouTube "Finding the Equation of a Straight Line", TuitionKit (5:57).

While we are on this subject, here is another thought experiment that is so simple and easy to do you can actually perform the gist of it yourself in a few minutes. But it's an experiment, if you were the one to do it, that could have made you one of the most famous and important people on Earth. That is, if you had managed to carry it out over 3000 years ago. So, just take yourself back in time and have a go at it. PS, you don't even have an intercept to worry about. Your straight line will pass through zero.

You are a scientist who existed in mid-1700 BC. The tools to do an easy experiment are available, but the knowledge to do it is missing. It is up to you to figure out everything you must learn and do for yourself. Performing this little experiment and doing it often enough will tell you, ***if you are smart enough,*** *exactly what the amazing experimental results that you are about to obtain mean. Having done this experiment over 3000 years ago you would have found yourself in today's record books counted amongst the greatest people who ever lived!*

Think of yourself holding a new tape-measure that you recently invented. (Imagine how valuable that measure would be today on Antiques Roadshow!) Your new tape-measure with its new numerical units, which you also had to invent, will allow you to accurately measure the spherically shaped coins of the day. Make two measurements, the diameter, D, of each coin and its circumference, C. As an aside, you have also invented the coordinate method of graphing before **Rene Descartes** *came up with it 3000 yrs. later (1596-1650, see "***Cartesian Coordinates", wikipedia.org***). With each coin so measured put the diameter of each coin on the horizontal x-axis and its corresponding circumference on the vertical y-axis. Then draw a vertical line up from each measurement on the horizontal x-axis and a horizontal line from each measurement made on the vertical line of the y-axis such that each line coming from the x-axis meets the corresponding line originating from the y-axis. Put a point at each of those intersections and draw a line starting from the zero-point, i.e. the origin of your graph's coordinates, so that your line passes adjacent to or maybe even on top of each point of the graph. If you have measured correctly, that line should pass through zero and on or very close to all of the points on your graph. And* ***because your experiment produced what appears to be a straight line****, you now know that you can use your new algebraic* ***formula of a straight line****, which you also happened to have just invented over 3000 years ago.*

Having used your new straight-line formula with the current experiment you replace your formula of a straight line, ***y=mx****, with* ***C=mD****. The slant of the line, which you have named* **slope** *and which you have represented by the letter* ***m,*** *is now easily calculated simply by dividing* **C** *by* ***D****, i.e.* ***C/D****. By the way, you also had to invent math and division, but that's another matter. You will see that the value of your slope is close to* ***3.14****. Again, it is fortunate that you are very bright because you would have also had to invent Arabic numerals, algebra and decimal fractions, but that too was no problem for you. So you focus on your experimental results and simply say, "That's very interesting".*

Remember, you still do not yet know what you have discovered. Therefore, you ponder, "I wonder if that number 3.14 will hold up and be true for larger circular objects? "Well heck!' you say, "That's easy enough. Let's do it again. I have a circular pond, a well, a bell, and a drinking glass. I'll just measure those and plot them the same way and see if I get another straight line."

You do the work and an hour later or so you have the results, which are now more accurate due to the larger size of the round objects and the fancy measuring device you invented with its ability to ultimately measure very large objects to six ***significant-figure accuracy****. What you find is that the number is the same and checks out but seems to be growing slightly larger as it has increased to a more accurate 3.142. "Hmm. That IS interesting!" Your curiosity grows. You are driven to measure all the round objects you can get your hands on, but after a long while you are starting to get frustrated. "What the heck is going on here?! The* ***irrationality*** *of these results is driving me crazy!" You know you are close to nailing it but you can't spill the beans to your friends or anybody else until you know for certain what the devil is going on. The problem is the value of 3.142 keeps getting very slightly larger as you measure larger and larger objects. Finally, driven completely mad by what you are finding you end up in a loony bin. Unbeknownst to you, you have also just discovered, for the first time in history, IRRATIONAL NUMBERS!*

Too bad, the world will never learn of any of your discoveries. Why is that too bad? Because what you have just discovered is ***pi****, one of the most important constants ever! But it turns out that it is a constant of a different color and it has driven you stark raving mad. You have been taken to the land of the totally insane because irrational numbers have no end. To this very day* ***nobody*** *knows the true value of pi. I just looked it up on Wikipedia and found the highly accurate value of* ***3.1415926535*** *is now followed by a million more digits and that number is still growing to an accuracy that has no end in sight! And is it a Universal Constant? It appears that that could be but it's an interesting question that apparently depends on things that I don't understand yet.*

Anyway, 3000 years ago you didn't stand a chance. Numbers that run off into infinity like this are so crazy and irrational that that is exactly what they are called, ***irrational numbers****. You discovered* ***pi****, one of the most important constants known to man or woman and you did it just by using a straight line and determining the line's slope. Instead of* ***C=mD****, you could have written C=* **πD***, that is, if you had invented the symbol for pi and realized it was an endless number. Having done this more than 3000 years ago you would have been way more than just a little famous today. Sadly, though, you ended up confined to an Egyptian asylum. Another Egyptian,* ***Ahmes in 1650 BC****, was very smart for his time but would have been no competition for you. He did not know the* *power of the slope on graphed coordinates. He simply estimated pi's number by matching circles with closely matched squares but that only got him a very distant close-to-the-right-answer.*

If Planck had the instruments that Einstein had when Einstein came up with his formula, **$E=mc^2$**, Planck might have come up with **$E=h\upsilon$** just by doing some relatively simple, empirical, *photoelectric-effect* experiments. Einstein conceived of the experiments but left others to carry them out to prove that light came in packets of energy or the "particles" called ***photons***. (An empirical experiment is like one that you just did that you could have used with math in an ***inductive process to show*** pi is a useful constant when it comes to anything to do with angular momentum). With Einstein's photoelectric experiment one can get a straight line by plotting the energy released from a zinc plate when photonic beams of light's wave/particles ***of sufficiently high frequency*** are directed at the zinc plate.

Zinc and other metals have "loose" electrons that are easily "kicked out of their loose-position orbit". The energy beyond which any such "kicking" can be done is precisely known as each metal's Work Function, W. The Kinetic Energy, K of the "kicked electrons" is what is actually measured in zinc-plate, photoelectric-effect experiments. That is, you will see **$E=nh\upsilon$** rewritten **$K=nh\upsilon-W$**. Increase brightness (n) or decrease wavelength (υ) and one should get a straight line with slope h.

The frequency of light on the *x*-axis vs the energy released on the *y*-axis will yield a straight line with a slope *m* that is equal to the Planck Constant, ***h***! That is, **$y=mx$** can simply be substituted with ***$E=h\upsilon$***. Amazing, isn't it? The following links will take you to two cool videos that explain and demonstrate the photoelectric effect.

YouTube "*Photoelectric Effect: History of Einstein's Revolutionary View of Light*", Kathy Loves Physics & History (16:33).

YouTube "*Demonstrate the Photoelectric Effect", Arbor Scientific*" (6:03).

By the way, Max Planck would have had a *much* harder time of it. He didn't have the photoelectric effect to give him ***$E=h\upsilon$***. He had to come up with ***$E=h\upsilon$*** from the equations of others who were befuddled by their complex equations that, unfortunately did not consistently agree with actual results at either end of light's extremes when light is eradiated from a **Black Body**. What the heck is a black body, you might ask?

YouTube "*Blackbody Radiation Oven*", Harvard Natural Sciences Lecture Demonstrations (3:25).

Thus, it wasn't at all straight forward or easy for Planck to rework the equations of **Rayleigh-Jeans** and **Wiens** into one that agreed with all wavelengths of light, emitting from a black body. So it's no wonder Planck is the one who got the Nobel Prize for he did just that. He ignored those equations and came up with his own, befuddling, Nobel Prize-Winning E=hυ!

YouTube "*Quantum Physics-Part1 Blackbody Radiation - Wien's Displacement Law", PhysicsOMG (9:59 min).*

YouTube, "Black Bodies and Planck Explained", PhysicsHigh (16:26).

Both are wonderful explanations of blackbody and what led to Planck's discovery.

One last thing, the fact that not all forms of light produce the photoelectric effect is the clue that Einstein needed to prove that light came in particle-like packets of energy. For example no matter how intense a beam of visible light one might use and no matter how many photons might be in that beam, visible light photons, which act like waves, as all photons do, are just not strong enough to cause electrons to jump out of their orbits. It takes ***only one photon of light*** of much shorter wave length and therefore, greater frequency, to do that. So let's forget that light is a particle for a minute and think about it being a wave. The distance from one wave crest to the next crest, or next trough to trough, is the wave's wavelength, **ʎ**. Now, think about it. If there is a long time to wait until the next wave attempts to drown our silly selves, it means we stand the chance of survival because the wavelength is long and the ***hertz*** of the wave is, as a consequence, a ***small number of wave assaults as calculated in cycles per second***. Also, a long wavelength wave is smeared out to a gradually rising, gradually decreasing swell. Abrupt smashing in a long wavelength would give us the experience of bobbing in a gentle swell.

However, if the wavelength is short, making the time between waves short, and the waves are still just as large, i.e., have a large amplitude, we will be blasted more frequently by raging surf and almost certainly drown due to the repeated pounding by the high energy, abrupt impacts that the giant waves bring down on us. Moral of the story, *it would have been better if we had not gone out so far.* Even though the wavelength of waves close to shore is the same as it is for the big waves, we wouldn't risk drowning because the energy of the sets is reduced because the ***amplitudes*** of the waves have been reduced to ripples due to the fact that we now stand in ankle deep water.

Now, let's go back to photons which act both as waves **and as particles**. Short wave lengths obviously lead to more waves per second than long wave lengths. So it should be obvious by now that at a given amplitude short wavelengths can deliver more energy per unit time than long wavelengths and the higher the wave, i.e., the greater the ***amplitude***, the higher will be the energy held by each wave. Therefore, if you want to knock electrons out of their orbits, a flashlight won't work because visible light waves have wavelengths that are too long with energies that are therefore too distributed due to the frequency of crests that arrive too infrequently? One might think something like that is going on if light always behaved like a wave. But it doesn't.

Contrary to the powerful effect held by the tiniest, high-frequency, short wavelength of UV-light, where even *just one photon* can jump an electron out of its orbit to ionize an atom, small, ankle-deep, ocean ripples no matter how many or frequent can't even cause tiny sand-granules to jump. That's one difference between waves and light waves because even one, short-wavelength, UV photon can dislodge a single electron.

And, here is another difference, and it's a big one when it comes to light. There is a threshold where light of long wavelength is too weak, *no matter its Amplitude*, to affect the orbits of electrons. That would be like giant ocean waves, no matter how big, having no effect. *Maybe you are thinking that it might be likened to a boat, far off in the ocean, gently bobbing about on the surface as a giant tsunami passes by unnoticed underneath the boat? Maybe, but let's not get carried away with our attempt to make the nano-world imitate the macro-world until we consider our other thought experiment.*

When it comes to light, visible light's wavelength is too long and because light can act like a particle, maybe we can think of long-wavelength light acting as weak, barbed, rotating, rubber-shots with energies that are too low to even perturb sand granules?

Both of the above analogies attempt to reconcile the fact that even if one uses light from a high-amplitude blaster-flashlight, where many photons deliver extraordinary brightness, electrons will not be moved from their orbits by such visible-light sources.

Instead, one needs a device delivering even shorter wavelengths that shorten to go way *beyond* the visible *blue.* That is, one needs **invisible-to-the-eye** light that includes ultraviolet, x-ray, and gamma-ray radiation. Light quanta of such spectra need only one photon to knock an electron out of orbit. And, to be sure, it's the photons with very short wavelength that are needed to create the **photoelectric effect.** *Now,* when you increase ***brightness***, you will increase the photoelectric effect as you increase ***amplitude***, ***A***, of short *wavelength-Λ*, photons. And when you increase frequency by progressively shortening wavelength you will have no trouble showing from the straight line that you get from your data that energy is proportional to ***frequency*** with a ***proportionality constant*** that is equal to ***h!***

But hold on. Are we not abandoning our thought experiment too soon? Maybe our tsunami analogy has some merit. Could the analogy of long-wavelength, high-amplitude tsunamis distributing their energy in the open sea to leave boats simply bobbing, as giant waves pass by, work? Could that analogy match what happens with long wavelength light? If photons arise from a field (see below for discussion of fields) where the field is a sea of writhing virtual photons, might not the uncertainty of the position of the real photons be increased, effectively diluted out by a surrounding sea of writhing virtual photons? Would this not leave the electrons simply bobbing instead of jumping? Is this not similar to giant tsunami waves leaving boats to simply bob when tsunami waves underneath a boat are diluted-out by a sea of writhing water molecules? Hmm? Did we not just come up with something potentially useful here using a thought experiment? Interesting, isn't it?

Light is certainly easy to talk about but it's really difficult to think about, is it not? Just for starters, we are being asked to think about the amplitude and wave length of a *particle* ! Well, I submit, as hard as any of us might try, some might get close to imagining it but in the end I can't imagine anybody will ever be able to accurately represent it. And yet that is exactly what I've been asking us to do here. I was about to give up and suggest we sell our shotguns and just get over it. But the tsunami analogy might have merit and so might the spinning shot. And then I also saw the following video. Maybe I spoke too soon. Maybe there is hope. Maybe we can visualize some of what light must be doing. I must say, I think you will agree even though you may have already seen it earlier, what you are about to see is a very cool, possibly enlightening macro-simulation of quantum reality.

YouTube "***Young's Interference Experiment with Single Photons Hamamatsu Photonics", Hamamatsu Photonics (10:45).***

Cliffhanger Question

Here is as good a place as any to attempt to resolve that Cliffhanger Question that I left for you above, i.e., "What is the speed of light in water?" My first thought was that the speed of light in water was slowed just because it was bumping into water molecules or something like that. A lot of people think like that, and maybe they are sort of right. But, I think, as the following videos show, although such a thing can happen, it is not the correct explanation for what is going on. And I'm still not sure that I understand it. You can let me know when you figure it out. Regardless, these three videos are fascinating.

YouTube ***"Why Does Light Really Bend?", The Science Asylum (9:31).***

YouTube "Why Does Light Slow Down in Water?", Fermilab (10:24).

YouTube "You Don't Know How Mirrors Work", The Science Asylum (12:11).

Also, for a brief read, **Quora.com, *"Does Light Actually Slow Down in a Medium" by* Viktor T. Toth.**

Paul Dirac (1902-1984), a great mathematician and physicist, created an equation that predicted space to be filled with seething, constantly colliding, virtual-matter and antimatter. In other words, as you may have seen in the earlier video on virtual particles, space is a colliding zoo of particles and antiparticles that render themselves into a virtual reality that is mostly invisible by constantly and instantly annihilating each other. Sounds crazy does it not? But, before one jumps to that conclusion, one needs to know that it isn't crazy at all and it may be the explanation for wave/particle duality of light and mass, that we seek. We actually routinely use the virtual world of antimatter wave/particles!

One example of active and valuable use of antimatter is the well-known ***PET Scan***, otherwise known to the medical community as ***Positron Emission Tomography.***. **And get this, that means that positrons are real and antimatter is real and it is not only real we play with it all the time in the form of these oppositely charged, positive electrons**! How? When positrons and electrons meet, they annihilate each other and produce a light photon that can be recorded in a PET scan.

YouTube "How Does a PET Scan Work?", NIBIB gov (1:33).

What? You can't be serious! I am truly shocked by this information. Are you not? I've always taken PET scans for granted. I had no idea we were playing with a seething space filled with a colliding zoo of particles and antiparticles. Did you? An entire Universe of antimatter could just as well exist as the one we are experiencing now. Our Universe could be instantly and completely annihilated by same and we'd never know it. Thank goodness for that last part. But the PET scan **is** hard-fact-proof for that possibility!

However, that's all beside the point. Of special importance to our current work and our attempt to understand the nature of light and the wave-matter or wave particles that assemble us, **Paul Dirac**, who we now know to be super smart, provided us with a beautiful equation that does a good job of explaining the source of Our Universe, and comes close to a theory of everything.

YouTube "*Atom: The Illusion of Reality*", Jim Al-Khalili / Science Documentary / Reel Truth Science (48:49).

Dirac also came up with a very useful analogy. He said something to the effect that we can't know how atoms and their electrons look, we can only know how atoms and their electrons behave. But relax, that's no problem since it is no different than it is with something with which we are all very familiar, chess pieces.

Chess pieces can look like all sorts of things. And what they look like does not matter, does it? It only matters what an identifiable chess piece does since that never changes. And the same is true for atoms, electrons and all other quantum particles. Therefore, even though my ping-pong ball models do not represent reality when it comes to electrons and their nuclei, I am on solid ground using ping-pong ball models to show 1. How atoms come to be, 2. How atoms can differ in their relative dimensions, and 3. How atoms work at chemical and biochemical levels. FYI, here is Dirac's exact quote: *"I can describe the situation by comparing it to the game of chess. In chess, we have various chessmen, kings, knights, pawns and so on. If you ask what a chessman is, the answer would be that it is a piece of wood, or a piece of ivory, or perhaps just a sign written on paper, or anything whatever. It does not matter. Each chessman has a characteristic way of moving and this is all that matters about it. The whole game of chess follows from this way of moving the various chessmen."*

Dirac's book on quantum mechanics is supposed to be one of the best. I have not yet looked at it. Although it is supposed to be good, I'm told it's not for the nonmathematical. You need linear and probably some Matrix Algebra to understand it. Nonetheless, I plan to tackle it, someday, maybe. **"The Principles of Quantum Mechanics"**: **Dirac, P.A.M**. (But for sure, see **Jim Al-Khalili's** excellent video production that describes Dirac's amazing mathematical discovery that changed everything that we thought we knew and, BTW, brought us the PET scan. I'll repeat Al-Khalili's YouTube link here just because I think it so important:

YouTube "*Atom: The Illusion of Reality, Jim Al-Khalili", Reel Truth Science Documentaries (48:49).*

And now let's blow the lid off everything that we've heard and seen thus far. The following three videos should do just that. The first one is a phenomenal.

YouTube "*QFT, What is the Universe Really Made of?", Arvin* Ash (14:57).

YouTube "The Baryogenesis Anomaly: What Happened to all the Antimatter?" (15:34 min).

I said at the beginning that I was excited to launch this new edition of OSAU with this Preface because I don't think any of us *really* know what's going on. Now you know at least one reason why I said that. The second video **above** deals with what **particle physicists** classify as **The fermions** and **The bosons** of **The Standard Model of Nuclear**

Physics. Light photons are massless, elementary force particles. They possess ***integer-spin*** and are, therefore, classed as bosons. Fermions on the other hand are wave-particles that possess what physicists characterize as ***half-integer spin.*** Without getting into futher detail, since photons are bosons they can occupy the same orbital space, and they ***can cohere*** to undergo superposition. And, since electrons are fermions with half-integral spin they ***normally can't cohere or*** occupy the same space and must act as good little leptons to obey the quantum rules outlined by the **Exclusion Principle of Wolfgang Pauli.** See two beautiful presentations below that are used to understand the full significance of the famous principle that was first introduced by Pauli to describe the behavior of electrons with the exclusion rules that were understood to be true in 1925 and are still held to be true today in 2022.

YouTube "*The Basic Math that Explains Why atoms are Arranged Like They Are: Pauli Exclusion Principle", Parth* G (10:36).

YouTube "*What Causes the Pauli Exclusion Principle?", Eugene* Khutoryansky (20:52).

But be prepared. These videos may further blow your already blown mind a bit. As for ***coherence***, that normally requires bosons. However, *alternatively, and this is what is important for quantum computing, fermions, nucleons and elementary quantum particles like electrons are not bosons, but they can actually be FORCED, at least for a very short time, into a very unusual willingness to be coherent and behave like bosons.*

As already stated, light-photons are members of the massless, force-particle boson group of the standard model of quantum mechanics. And like all bosons light-photons can not only share space with another photon, they can **superposition** with other photons to become **entangled** and act in total **coherence** to join in that space as an identity with their neighbors. And it is this boson property of coherence that allows light to be able to perform as a **laser.** In addition, if I have it right, it's ***coherence*** that will **give us quantum computing** by way of a process that involves manipulating **atoms (i.e., fermions) under conditions of extreme cold to behave like cohering bosons!**

YouTube "*Bose Einstein Condensate Coldest Place in the Universe*" (6:12),

YouTube "*Quantum computers Explained with Quantum Physics",* Quanta Magazine (9:59).

YouTube "*How Lasers Work – A Complete Guide*", Scientized (20:45).

Here we are, nearly to the end of this Preface. But there is still a lot to answer for when it comes to that big question, "How the heck do we know this stuff?" You now should have a pretty good idea about why we think we know some of the things described herein but you have also just seen that "we" are still nowhere near done trying to figure things out. For example, you may have thought I was doing okay at presenting the evidence for our understanding of the existence of atoms, quantum mechanics, the properties of atoms, etc., and especially how the nanoscale world has so miraculously been assembling everything from day one.

But then, did I not just say "QFT!" And, FYI, that's not a swear word though it might sound like one if one tries to pronounce it as a voiceless-stop that comes out sounding something like QuiPhuTa. Quantum Field Theory (QFT) and Quantum Electrodynamics (QED) are worth a look. *Arvin Ash is wonderful and his YouTube,* ***Mix – Arvin Ash****, is the* ***best*** *mix I've seen on this subject.* From the mix, I specifically recommend the intro to QFT and QED,

YouTube "*Arvin Ash, Are Photons and Electrons Particles or Waves? Make Up Your Mind God!", Arvin Ash (14:45).*

Also, if you are interested to learn more about QED, the following presentation on **Feynman Diagrams** is excellent.

YouTube "*Can I Explain Feynman Diagrams in 60 Seconds? #shorts", DoS-Domain of Science (0:60).*

And one can actually see Feynman's amazing brain in action. This is great.

YouTube "Ri*chard Feynman. Why", firewalker (7:33).*

YouTube *"The Complete Fun to Imagine with Richard Feynman"* Christopher Sykes (1:06:50).

There are countless excellent videos and audios with Feynman talking about quantum mechanics. These videos are fun to see and listen to when you realize that you have been given the chance to be in the virtual presence of a great man, the famous father of QED.

Indeed, you might ask, "What is all this talk about "fields" that you just snuck in at the end of this Preface?" To tell you the truth, as much as Our Universe seems to have a lot to do with waves, it likely has more to do with the "fields" in which the waves find themselves and, unfortunately, those *field* things are still way beyond my paygrade. Why so? Fields deal with the vector and tensor mathematics that one can see in the old pictures of Einstein's legendary, equation-covered blackboards. And it was, specifically, the more difficult of these two, tensor mathematics, which Einstein had to use for his discovery of what may be the world's most important mathematical revelation, The Theory of General Relativity. Also, fields were at the root of Paul Dirac's ground breaking formula and mathematical insight into Our Universe's seething invisible world of empty space filled with virtual particles and antiparticles.

It's interesting, isn't it? Paul Dirac's quantum field that I call his virtual particle/antiparticle, seething sea-of-uncertainty led to the notion which came from the thought-experiment above, that the tsunami analogy might help explain light's cutoff. While I was writing an example of the thought experiment that spoke of water molecules diluting a tsunami to leave a boat simply bobbing at sea when one of the giant waves passed by harmlessly underneath a boat, the idea that Dirac's sea-of-uncertainty, could create a similar diluting effect on long-wavelength photons came to me. Could it be that this effect explains light's cutoff when a visible, relatively long-wavelength version of it is used in photoelectric-effect experiments? Maybe I'm crazy but let's see if the idea turns out to have merit. Fortunately for you, I will be the only one to be tarred and feathered if the idea is totally bonkers. But if it's not, you should know, dear readers, that I regard you as accomplices. I only came up with the idea in my attempt to demonstrate how a thought-experiment might work. It's certainly a "Holy Moly!" moment if this thought experiment actually comes up with something useful. I'd not be surprised if there weren't better explanations but I've yet to find any other than the obvious explanation that photons with long wavelengths just don't have the energy to knock an electron out of orbit. At least my search for a better explanation wasn't wasted. Some other questions came up.

1. Does Dirac's equation support the idea that there are diluting-effects that arise from quantum-level uncertainty when photons interact with Dirac's seething, virtual, particle/antiparticle, quantum-fields or, as referred to here, seas?

2. Does the math behind such a hypothesis support and explain the cutoff? I won't reproduce Dirac's equation here but one can see it again in Al Khalil's video that I have reposted below. I'm no mathematician, but if I'm to continue with this conjecture it's clear that I need to find one or bone up on the math myself. Hey, maybe one of you already figured this thing out. No matter, I plan to work on understanding Dirac's equation, even if I have to learn tensors to do it.

3. Has anybody tried to build an x-ray or gamma-ray laser? Answer, yes. See below.

4. It seems to me that light might be losing some of its wavelike properties at the cutoff to become more shotgun, particle-like. Do the particle-like properties of photons continue to increase as photon-wavelengths continue to decrease? There clearly is one particle-like property that light does not lose, E=nhυ. On the other hand, there is also one wave-like property that light does not lose. Light IS always an electromagnetic wave!

5. Did you see the gamma-ray laser videos I've posted below? Yikes!

YouTube *"Atom: The Illusion of Reality", Jim Al-Khalili / Science Documentary / Reel Truth Science* (48:49).

YouTube *"Introduction to Tensors", Faculty of Khan* (11:14).

YouTube *"This powerful X-ray Laser Can See the Invisible World", Seeker* (8:27).

YouTube *"Gamma-Ray Laser Moves a Step Closer to Reality", Research Info* (2:27).

YouTube *"Introduction to Tensors / Faculty of Khan",* (11:14).

Back to Quantum Field Theory. As far as QFT is concerned, Richard Feynman was able to carve out an important way to understand quantum fields with his diagrams and Quantum Electrodynamics. Anyway, if it hadn't been for PET scans I'd probably not have mentioned fields here at all. And that would have been a big mistake because fields appear to be at the crux of the matter. And, really, what could be more interesting than an invisible universe of empty space filled with virtual particles that constantly seethe and annihilate each other? It does get one thinking and not wanting to give up one's efforts to understand the un-understandable a bit more. So, have a go at it with Maxwell's equations.

YouTube *"Divergence and Curl. The Language of Maxwell's Equations, Fluid Flow, and More", 3Blue1Brown* (15.42),

As for me, I'm still enthralled by Planck's ***h*** *and the way that constant generates units of* ***Planck-Time*** *and* ***Planck-Length, the unimaginable minimums of space and time*** *that characterize the earliest moments of the Big Bang!*

YouTube *"What the Heck are Planck Units?", The Science Asylum (7:04).*

Based on these apparent facts and our continuing conundrums over which we've gotten ourselves all in twist in our feeble attempt to understand light, I think it is safe to say that whatever anyone thinks or believes to be true in 2022 might be wrong in 2023. No doubt, more shall be revealed. That fact, strangely enough, gives me a bit of solace now that I know I'm not alone in my attempt to understand the un-understandable. But before any of us get to cocky have a look at one more video:

YouTube *"The Mandelbrot Set – The Only Video You Need to See!", The BitK (21:18).*

PS, I think you and I might have a lot of time left to try to figure things out in Our un-understandable Universe. Did I not tell you? I'm planning to move to Mars where I will live on a magnificent view-lot for at least 100 years. Thereafter, I'm not so sure, but I am sure about one thing, even then, there will be much more to be revealed.

Upon returning to "reality"? I shake my head, *"You say that I am some kind of strange animal being? And then you tell me my sequenced DNA doesn't make me a whole lot different than a plant?! And you go on to tell me that I am trapped with a whole lot of other strange beings, which have DNA that's much the same as mine, living on the surface of a spinning ball that's whipping through space at 1,300,000 mph with 7, 000,000,000,000,000,000,000,000,000 atoms re-assembling me each and every minute of each and every day as I travel?!* ***Holy Higgs-Field, Batman! YOU HAVE GOT TO BE KIDDING ME!!"***

If you still find yourself, as confused as I am, asking the same questions; Who am I? What am I? Where am I? Why am I? When am I? The following, fun to watch, beautifully clarifying **YouTube Films** by **Jim Al-Khalili** may help. But, don't miss the first Al-Khalili video rescued on YouTube by Abubakar Hamid. It just might be the most important one because it is a beautiful development of the science of light from Euclid to Galileo, Hook, Newton, Maxwell, etc. For example, Al-Khalili's description of Newton's trip to a Fair in the 17th century should help answer many of your questions that you must now be having. For example, *"How the heck do we know how stars work? How do we know how old stars are? How do we know how far away they are? How do we know how many stars there are in the universe? How do we know how atoms assembled themselves? How do we know how old Our Universe is?"*

YouTube *"Light and Dark 1 of 2 – Jim Al-Khalili", Abubakar Hamid* (59:33).

YouTube *"The Story of Energy with Professor Jim Al-Khalili",* Spark (59:00).

YouTube *"How Information Helps Us Understand the Fabric of Reality, Order and Disorder", Spark* (58.46).

YouTube *"Beyond the Atom, What Really is Everything?", History of the Universe (42.59).*

YouTube *"Atom: The Illusion of Reality, Jim Al-Khalili", Science Documentaries by Reel Truth Science* (48.50),

YouTube *"The Story of Electricity Full Episode, Jim Al-Khalili",* Team Aphrodite (2:56:15).

"The Illusion of Reality" reminded me of my promise to discuss Einstein's versions of reality that were based on his thought experiments. Think about our "existing" Universe expanding into nonexistence. That's like "Time" from Einstein's relativistic "Space/Time" expanding into NoSpace/NoTime.

As you stand at one end of a coach on a train, throw a ball that clocks at 60 mph. An observer outside sees you throw the ball as the train speeds by at 60 mph and they clock the relative speed of your throw at 120 mph. Now pretend you are on a light beam throwing that ball. **Researchgate.net "Chasing the Light: Einstein's Most Famous Thought Experiment", John D. Norton.**

As you stand in an elevator that falls through space, you notice that you are weightless. You also notice that light penetrating a small hole in the elevator shines VERY slightly elevated on the other side of the elevator. In other words, the light beam appears to be bending VERY SLIGHTLY upward (light travels VERY FAST) and you appear to be weightless because the elevator is falling. Stemming from the first type of thought experiments came Einstein's Special Theory of Relativity, Space/Time Unification, and the conclusion that travel at 186,000 miles per second (the c in $E=mc^2$) would let one live the impossible dream, i.e., forever at infinite mass. Stemming from the second type of thought experiments comes Einstein's General Theory of Relativity. The identity of gravity with the action of falling in an elevator was one such experiment with an illustration of the gravitational bending of light as an added bonus. Strange as they all may seem to be, these theories have all been proven correct by experimental evidence.

Researchgate.net "Chasing the Light: Einstein's Most Famous Thought Experiment", John D. Norton.

YouTube *"Einstein's Greatest Legacy: Thought Experiments", Sabine Hossenfeider* (7:27).

YouTube *"Special Relativity simplified Using No Math. Einstein Thought Experiments", Arvin Ash* (12:18).

And you might ask, what about your promise to reveal more about your idea for w h a t you t h i n k t o be a beautiful equation, C=I=L=E=mc2? And "How the heck do we know that?" If I were you, I'd ask Brian Green and Max Tegmark. Also, if you ever wondered if we were at the center of Our Universe, be prepared, maybe we actually are!

YouTube *"Our Mathematical Universe: Brian Greene & Max Tegmark", World Science Festival* (1:36:57).

YouTube *"Axis of Evil Max Tegmark", Black Swan Thinking.* (19:30).

YouTube *"The principle documentary", Paulo Roberto* (1:28.09).

Sheesh! I almost forgot, herein follows the edited version of the book that I wrote in 2015 and is now named ***"Our Self Assembling Universe-2 (OSAU-2)"***. Which is also now an **AWTbook™, error-free, ping-pong-ball-facilitated, personal, tactile experience with atomic nuclei.** I anticipate that this new **AWTbook™** form of presentation will forward one's understanding of the bewildering subjects just presented. At least that is what I hope, ***but I hope even more, if you happen to be a teacher, that your pupils will enjoy Our Self-Assembling Universe 2022 in its new AWTbook™ format.***

Addendum

"Forcefield" as defined in OSAU-2

Forcefield – *An illusion of apparent structure*. The structure can be easy to turn on and off, as it is for the force of an electromagnet, or as it is exclusively represented by the ***yet-to-be-invented*** forcefields that are often represented in Science Fiction. Otherwise, they are difficult to turn on and off as represented by ourselves and the structures of our daily lives. The latter forcefields can be visible or invisible, colored or uncolored, hard or soft, smooth or rough, hot or cold, dry or wet, noisy or silent, animate or inanimate, alive or dead, conscious or unconscious, unknowably vast or inconceivably tiny, seemingly real or obviously unreal. *To be clear, examples of the Science Fiction variety do not yet exist. Most of the examples that do exist,* are well-known to our daily lives as they include our very own bodies, water, rocks, insects, plants, animals, planets, stars, galaxies, and all other things of substance including atoms and all the things that the atoms self-assemble. This definition of the compound word, *forcefield*, makes it clear that all seemingly solid objects are actually nothing but illusions masquerading as mass. Most all examples of mass that we encounter in our everyday lives consist of smaller forcefields called molecules, which are assembled by yet smaller forcefields called atoms. All forcefields smaller than atoms are members of the **Standard Model of Particle Physics. Atoms** are forcefields assembled by the electrons, protons and neutrons of the Standard Model. The standard model is divided into the three major categories; **Bosons, Hadrons,** and **Fermions**. The forcefields called **Protons, Neutrons** and **Electrons** are the **Fermions**. All three are forcefields that act as particles of mass. However, the **Proton** and **Neutron Nucleons** are further classified as **Baryons** because they are made up of elementary, half-spin particles called **Quarks**. **Electrons** on the other hand are, themselves, elementary particles that have been classified by physicists as **Leptons**. Quarks are elementary particles in the Category called **Hadrons**. Light as we now know it, consists of elementary particle forcefields called **photons** that find themselves classed as massless, **Boson Force-Particles** along with the now famous **Higgs Boson** of the Standard Model. All the particles of the Standard Model are, themselves, forcefields that have been manifested by vectored-fields-of-force that writhe within Paul Dirac's seething sea of virtual, particle/antiparticle uncertainty. Thus, all of the above can be de-materialized as fields of the E in **E=mc2**. Or, if you like it, without my knowing the relevant units, "materialized" as the C, I, and L from the E in **E=mc2**. If you have heard of but don't know what to make of all the excitement over the Higgs Boson, the following video does a great job of explaining it, how it was measured and how the Higgs Boson fits into the Standard Model of Particle Physics. It's also perfect for sit-down *viewlistens* and physically active *justlistens* as one *walkjogruns* or *weightlifts*. I've already done both myself with this particular video twice.

YouTube *"Beyond the Higgs: What's Next for the LHC? – Harry Cliff"*, The Royal Institution (59:44).

Our Self-Assembling Universe-2

2022

A Ping-Pong-Ball-Facilitated AWTbook™

For the Curious of All Ages

One morning, out for my morning run, I had a revelation. Taken aback by the beauty that lay before me, I heard myself uttering in amazement, "What a marvelous, ingenious invention!" That involuntary response, and a sudden inexplicable change in perspective, left me feeling like a recently-arrived, body-snatching, interplanetary space-traveler. I could feel my new planet rotate on its axis. I could see in detail its vast collection of water molecules crash onto a beach of pulverized silicon dioxide amalgams. I could feel photons exciting my newly sensitized retina, and optic nerves sending signals to my newly enhanced image-intensifying brain. I could see, therein, strange beings standing by a seawall, where, with my newly crafted ear drums wildly vibrating and tiny calcium-capped hairs excitedly waving signals to auditory nerves, I could hear one of the beings demanding attention with a one-word command, "rainbow!" And, then, as if I had awakened from a hypnotized trance to receive a posthypnotic suggestion, I finally got it! "*Everything in our universe has been and continues to be* self-*assembled! Atoms, self-made by electrons, protons and neutrons, working together, on their own, have made, and continue to make, everything!"*

Assembling the Universe Atom by Atom

A widely accepted cosmological hypothesis holds that our universe started in a "Big-Bang" way from subatomic particles and atoms about 13.8 billion years ago. The infinitesimal point where it all started has been called a "singularity" which is, as I think of it, just another way of saying, "Huh?! You've got to be kidding me!"

Theories concerning the singularity and the Big-Bang are fascinating but not needed to appreciate the fact that every day we are witness to an incredible magic show where atoms, as we speak, self-assemble themselves into amazing illusions of mass, chemistry, and structure!

Having finally absorbed this now obvious yet profound truth, I wanted to find a way to see and experience it for myself. I didn't know how I was going to do it but something led me to a drugstore for the answer. Why there of all places? I don't know, but I felt guided. Somehow I *knew* I would find ***it*** there. I carefully searched every department in the place; games, photos, food, paper-goods, clothes, art, auto supplies, and even the stuff a drugstore is supposed to carry, pharmaceuticals. I also asked for help but, not surprisingly, all I got were placating, nervous, quizzical looks from customers and staff. I was about to give up looking when there *it* was! Hanging from a shelf, lined with various brands of deodorant, were packages, containing six each, of orange and white ping-pong balls. "Perfect", I said to myself. "Just what I *didn't* know I was looking for!"

The strange events leading to the revelation, the compelling urge to search a drugstore for a way to *experience* it, and the finding of the perfect atomic models for it hanging from a shelf with underarm deodorant, left me in wonderment and with a certainty that I had, indeed, been guided and handed a mission. I was to get to know, on an intimate level, the atoms that are, even now, assembling me, assembling you, and self-assembling the entirety of the amazing universe* of which, incomprehensibly, we all exist as an integral part.

*I often refer to "our" universe to leave open the mind-blowing likelihood that there are more, maybe *a lot* more. Enter "multiverse" into an internet search field. Especially, take a look at what **Wikipedia** and **YouTube** have to say about it.

YouTube "*The Multiverse Hypothesis Explained by Neil deGrasse", Science Time (10;02).*

The Drugstore Deodorant Isle.

Atoms, the Self-Assembling Building Blocks of our Universe

Most of us know that we are made of atoms. And most of us have some idea about the makeup of our flesh and bones. For example, we know that our bones contain calcium, Ca-40. We know, if we don't get that atom in our diet, we will get sick with rickets and die. We know we need water, H_2O. So we also know we need hydrogen, H-1 and oxygen O-16. We know our flesh is made of protein. And even though it's likely only a few of us know details of the chemical structure of proteins, most of us know that carbon, C-12, and nitrogen, N-14, are important components. We know helium, He-4, from party balloons and lithium, Li-7 from drugs and storage batteries. And we know about aluminum, Al-27, and iron, Fe-56, from the metal atoms found in our cars and buildings. And we know about silver, Ag-107, and gold, Au-197, two of the metal elements that used to be common in our coin currency.

So, there's no argument, right? Most of us know, and do so with a high degree of confidence, that we, our planet, our solar system, our galaxy, and all other seemingly solid objects in our universe, both unimaginably large and invisibly small, are made of atoms. And yet, even though I have said it many times and thought it many more times, I hadn't truly assimilated the very obvious fact that our universe isn't just made *of* atoms, it has been, and continues to be, self-assembled ***by*** them. Atoms have been putting the universe together, and they, and the very forces that made *them*, have been doing it all, on their own, subatomic particle by subatomic particle, and atom by atom, from day one!

I believe the Big-Bang cosmologists have it right, but I think I can safely say that no one, at least not on our planet, has any idea what may have happened before the Big Bang, unless it might be Brian Green, or even really knows anything about the very early-early Planck Time events that led up to the Big-Bang. And, although the evidence is quite strong for the Big Bang, there may be some very few scientists who still think the universe came about some other way. But most scientists, who theorize about the early beginnings of our universe, talk about an energy, or more specifically a soon to be vectored energy, thought of these days as a field of force, came into being from nothing, or at least practically nothing, almost precisely13.8 billion years ago. And, nearly at the same instant, that field of force divided itself into four parts, i.e., the *gravitational*, *electromagnetic*, *weak*, and *strong* forces that are the drivers of the standard model of particle physics.

YouTube "***Why & How do the 4 fundamental forces of nature work?, Arvin Ash"* **(15:32).**

The same theorists posit that the four forces somehow *worked together* to assemble, in a Big-Bang magic instant, all of the quarks, photons, electrons, protons, neutrons, neutrinos, and other particles & anti-particles that make up our universe. Just enter ***Big-Bang Timeline*** into an internet search-field or go to **YouTube** for astonishing Planck space/time accounts. Try talking to your phone like I do. **"Hey! Siri, YouTube, Big-Bang Timeline"**. Or get the same result in a different format by leaving out **"YouTube"** with just, **"Hey Siri, Big-Bang Timeline"**. Alexa or some other AI should work as well. You might also enjoy watching the following using the same type of voice command.

YouTube *"Timelapse of the entire Universe", melodysheep* (10:49).

YouTube *"Timeline of Universe / Big Bang to Today", Astrogeekz, (5:06).*

YouTube *"What Was the Universe Like Immediately After The Big Bang?", History of the Universe (29:02).*

And the newest thing on the horizon may very soon let us ACTUALLY see the Big Bang happen almost form the beginning. ***It may also prove that we are not alone in Our Universe!***

YouTube *"How NASA's Webb Telescope will Transform Our Place in the Universe", Quanta Magazine (14:54).*

The launch of the telescope is just a few days ago Dec. 22, 2021. And what about life? What may life look like elsewhere in our Universe?

YouTube "Life Beyond, Chapter 1. Alien Life, Deep Time and Our Place

in Cosmic History 4K", melodysheep (30:26).
Suffice it to say, according to current theory, our universe, within its first second to three minutes of its arrival, *created* all of the proton and neutron building blocks that were needed to assemble the first four atomic nuclei: hydrogen, helium, a little lithium, and just maybe a tiny amount of beryllium. And then, in less than 20 of the first minutes, having accomplished this miraculous bit of self-assembly, all the atom-assembling activity came to a screeching halt followed by a hiatus that lasted for nearly 300 million-years!

I hope you won't mind my contrivance, but I am going to leave the explanation for this mysterious event hanging in suspense, akin to a soap-opera cliff-hanger, whilst I air out some matters of importance.

As far as the present cosmological subject is concerned, I consider myself to be a new student, one who has just begun to learn how atoms self-assemble things. Even though I've had a long academic and commercially successful scientific career and have had the opportunity to teach and do research in microbiology, biochemistry, enzymology, protein chemistry and molecular genetics, I only became aware enough myself in 2015 to want to learn more about atoms. Suddenly, I needed to know the specifics of how electrons, protons and neutrons self-assemble atoms, and how those atoms go on to self-assemble all the other "hard" stuff in our universe.

As of April, 2015, I had learned a lot, mostly from the internet, especially from the websites of Wikipedia and YouTube, but my understanding of the beginning of our universe and its assembly by atoms was then and still is very much a work in progress. And while this admission might seem to make me unqualified to write about such an important subject, as I explained at the opening, *I have been guided by a rainbow and the very atoms that, even now, as I write this, re-assemble me*. So, kindly, blame the impatience of atoms for my impudence and accept this re-work of my 2015 opus that I just hope will be somewhat entertaining and useful.

It's from that entertaining and useful perspective that I am very excited to present the republication of Our Self-Assembling Universe as an AWTbook™, a new, evergreen-type book that's laced with powerfully relevant **A**udio, **W**ebsite and "**T**ube" references. This new book-form, as already described in the Preface and reiterated here, include the author's choices for the relevant, best, **A**udiobook, **W**ikipedia and YouTube titles which are, in each case, carefully selected and noted in the following format; **YouTube "*Title",***

Author/Source (duration). AWTbook™ links can be voice-activated, as they are herein, or they can have links that are activated by some other means, such as the newly prevalent photo-activated QR links. Or if published as an E-book, activation can be accomplished with hyperlinks. As for this, my first AWTbook™, I have found specific title-entries, when entered into one of our ever-improving, artificially-intelligent, mobile-phone or tablet search-engines manually or by voice command to be an effortless, instantaneous and satisfactory active reference mode. Also, if all else fails, one can always go to the given website and type the given title into that website's search-field.

Even though, at this writing, my AWTbook™ is the only such book in existence, I can unreservedly vouch for the following. With one's cellphone, tablet, computer or OLED Smart-TV close at hand to be used as one reads, one can realize, possibly for the first time, their most exciting, informative and satisfying experience upon reading a complicated, difficult-to-understand, technical book.

Moreover, one can find the AWTbook™ to be a great way to keep all the information that one learns relevant and instantly updated with new titles that reveal themselves as one browses the selected titles. I've also found by writing this book that the act itself of writing such a book, whether it gets published or not, can be an excellent way to teach oneself a difficult subject as one creates their own personal Reference-AWTbook™ by adding as one goes YouTube and other activate-able sources.

And now, lest I forget, I'd like to draw your attention to my rules for assembling the universe with ping-pong balls. They are just that, my rules. Aside from accurate accounting of the protons and neutrons in each atomic nucleus, the white and orange representatives that I discovered in a drugstore are arranged in my models of atomic nuclei solely at my discretion.

In addition, as will be discussed, it is physically impossible for my ping-pong ball atomic nuclei and atom models to accurately represent the real things, especially since the real things are *all* nothing but wave-particle *forcefields*, i.e., things that are almost impossible to think about in any concrete terms. Even so, I think my models will prove to be very useful, especially since they can encompass the entire periodic table. And, I believe that my models will help me, and others who follow my developing treatise, to acquire an intimate, appreciative view of the miraculous self-assembling nature of our universe.

The Schrodinger Equation combined with modern, computational, virtual-reality transformations has given us accurate, quantum mechanical views of the probability zones for the odds of finding a given electron in its orbital. This is as close to reality as we can get to view the actual state of affairs for the electron clouds of atomic orbitals. The following videos are the best I've seen for explaining what we think is going on in this realm. What is going on in the insides of an atom's nucleus, as represented by ping-pong balls which scale at about 100,000 times smaller than an atom with its full complement of electrons, is, I'm afraid to say, *an entirely different matter.*

YouTube "*Quantum Mechanics and the Schrodinger Equation*", Professor Dave (6:28).

YouTube "*Atomic Orbitals Visualized Dynamically*", Science Asylum (8:39).

Ping-Pong Ball Assembly-Directions& Rules

1. Assign white balls to neutrons and orange balls to protons; 2. Assign the correct count of orange and white balls for the isotope most representative for each new atomic nucleus (some nuclei, as you will learn, have more than one naturally occurring isotope, i.e., nuclear versions that contain different neutron counts. (Our models will only represent a major version); 3. Separate the orange balls with intervening white balls; 4. Conserve space as much as possible by placing the balls within a space-filling, spherical shape by fusing the balls together with fabric tape.

Except for an off-the-wall idea, as described below for the role of neutrons and their specific arrangement in *neutron cores*, that's about all there is to know about my rules for constructing models of atomic nuclei. But, it is important to note and constantly keep in mind, nobody will ever be able to make accurate physical models or represent on a sheet of paper the *nuclei* of atoms, and certainly, there is no way to even come close when it comes to the atoms themselves. That said, even though nobody can accurately model the nuclei of atoms, there are ways to create useful representations so that one has something to hang one's hat on to facilitate visual imaginative thought. And that said, the way I like to visualize atoms and their nuclei is as follow.

Even if we were able to *see* a model of an ***atom*** magnified to the point that its constituent ***baryons*** (aka, the proton and neutron ***nucleons***) were the size of ping-pong balls, we wouldn't be able to see those baryons. Instead, all we could possibly see at the diminishing perimeter of the ***atom's*** electromagnetic "forcefield"* is what would appear to be empty space. If you have ever experienced the repelling force of the like-poles of two magnets, you will know what I mean by an invisible force.

** Sorry for the interruption but I need to clear up something Fields of Vectored Force in physics are one thing. Force-Fields or the forcefields of Science Fiction are another. I like the compound word-construction, **forcefield**, even though that compound word has yet to appear in Webster's Unabridged Dictionary or as a legitimate SCRABBLE® word. Regardless, I will use that construction here. Maybe the compound word-construction eventually will be recognized by the definition described herein. In my opinion the terms, Fields of Force, Force-Fields and forcefields, desperately need to be understood, distinguished, defined and used appropriately.*

*A forcefield in Science Fiction (Sci-Fi) is simply an intentionally generated region of semi-impenetrable space. The invisible force that one experiences with opposing magnets is an instructive example. But knowing now what we think we know about atoms, I think the term, forcefield, is also a useful word to depict the illusion in which we have found ourselves wherein quantum-state structures masquerade as solid-mass. Therefore, the main difference between the forcefields of Science Fiction (see early examples in William Hope Hodgson, "The Night Land", 1912, John Campbell, "Islands of Space", 1931), and the forcefields to which I refer, are that the latter are the forcefields of which we are most familiar but normally don't think of as such. It's just that unlike the invisible forcefields of Science Fiction, Out Material World forcefields are mostly quite visible, colorful illusions of structured hard-stuff that are generally difficult to turn on and off. Objects in our material world **can** be turned on and off, but turning them back on might take some ingenuity, time and a lot of added energy to get the job done.*

In other words, what I am proposing is that we make the invisible, fictional forcefields of Sci-Fi equivalent to the visible, real-life forcefields that produce the illusions of mass in our everyday existence. And let's define Vectored Fields-of-Force to be theoretical fields from which the forcefields of our everyday lives are thought to be derived. This distinction should make it clear that our world of apparent hard-stuff is, indeed, an illusion consisting of nothing more than vast assortments of forcefields of different quantity, quality and function. These forcefields, by my definition, are visible, tangible examples that otherwise closely match the less visible and less tangible forcefields first imagined by the early authors of science fiction. However, without quantum mechanics to back up their understanding these early authors would not have had the clues needed to allow them to conceive of the additional, applicable and very useful usage of their term as described here.

Something just occurred to me as I was writing the above forcefield-clarification. The concept, as I am presenting it is new enough to be written as a new Webster, Scrabble-Dictionary-Acceptable word where the above definition relates to the real things and beings of our everyday lives as well as to the proposed forcefield invention as described by Science Fiction. Atoms are not solid objects. They are illusory particles and all of the structures they assemble are illusions. Everything with which we interact are forcefields conjuring the illusion of solid objects. Everything that one can see and touch, including your very self, is an illusion created by the forcefields that we call atoms. I think we should claim the compound-word "***forcefield"*** for these illusory things**.** Although, this form of the wording for a science-fiction forcefield is not new, I think my use of the word ***forcefield*** as a new compound word (surprisingly, heretofore, an unrecognized English word-construction) should now be included to describe the world around us **as it truly is**. Others have already used the term in its unrecognized, compound word-form so let's make it legal and adopt it. The following video is a good example of that use and a recent representation of science-fiction's use of the term.

YouTube, ***Are Forcefields Possible / Up and Atom", (8:08***).

To summarize: Rocks, glass, wood, metal, human flesh, etc., are all just one manifestation after another of ***macro-forcefields*** *assembled by atomic,* ***nanoscale-forcefields****. And these forcefields differ from the original science-fiction version only by the degrees of difficulty associated with manifesting, penetrating, and switching. For example, it's hard to manifest atoms. It took a Big Bang to do that. It's hard but, apparently not impossible, to manifest a science-fiction forcefield. We just have yet to manifest one. It was hard to manifest an atomic bomb but we finally manifested one on July 16, 1946, whereby, we instantly switched it off by blowing it up. It can be hard to penetrate all forcefields but all of them can be penetrated depending on circumstances. If one had neutrino-receptive-eyes, one could unsee our entire planet disappear to become nearly invisible as our entire planet is penetrated by almost all neutrinos.*

YouTube *"Ask symmetry – Why can a neutrino pass through solid objects?", symmetry magazine (3:37).*

But of all these forcefields there is only one type that can be manifested in a form that is easy to switch on and off. That forcefield is the forcefield of science fiction and, astonishingly to me, it is a forcefield that has yet to be realized, at least not yet on this planet. You get the idea. With the forcefield definition so clarified, it should be clear that it is the Vectored ,Quantum-Fields of Force that are responsible for manifesting them all.

The forcefields that we call atoms, which were manifested by the actions taken early on in the origins of Our Universe, have in turn, self-manifested, i.e., self-assembled, all of the microscopic and macroscopic forcefield entities in Our Universe. And that includes you, me and the very chairs on which we sit. An obvious example of the type of forcefield to which I refer is the orange ping-pong ball. The surface of the ball is the extent of the main body of its forcefield which is in turn exerted by a large collection of atoms, some of which are even able to absorb blue light to reflect by their collective force a blend of wavelengths that we call the color orange. The surface of the spherical-shaped forcefield is also strong enough for us to grasp and bash with a strength great enough to resist physical deformation. Alternatively, a Sci-Fi forcefield ping-pong ball or chair that one could turn on and off, could be made to be even more invisible than a pane of glass. In fact one may find such a forcefield chair in your home on Mars. Can't you just see yourself sitting on an invisible chair gazing through an invisible wall to a future become reality?

Back to the Ranch. Once again, I'm sorry for the interruption. However, I think that interruption could turn out to be quite important.

So, even if you could see the perimeter of the atom's electromagnetic forcefield, the comparative enormity of this see-through, spherical thing would make your model atom so huge its nucleus would be so far away that its forcefield of baryons, even though they be the size of ping-pong balls, and even though you had magic eyes that could see them, would still be hopelessly invisible with those baryons residing more than a half-mile away!

YouTube "The size of the Atomic Nucleus", Andrew Reader (3:39).

And I repeat, lest anyone think otherwise, anything that behaves simultaneously as both a wave and a particle, or a particle managed by mathematical wave functions, is I believe, beyond anyone's ability to actually visualize.

None-the-less, let us not despair. There is a real-life, macroscopic, possible-look-a-like representation of a wave-particle that might help clarify this confusing matter. It is a water-droplet and a **Louis de Broglie**-like pilot-wave. The "water-droplet effect" is not made up. It is a real, macroscopically observable phenomenon. And if **Paul Dirac** is right, I think something like this could be occurring with **virtual particles** in the fabric of Dirac's seething field of empty space.

YouTube "*Is this what Quantum Mechanics Looks Like?", Derek Muller (7:41 min).*

Also, at the outset, I think it is important to fix in one's mind the fact that 20 minutes after the birth of our universe, no true ***atoms*** existed, instead only three, and maybe just a tiny bit of four, types of atomic ***nuclei*** had thus-far been assembled, those of hydrogen, helium, lithium, and an unlikely tiny bit of beryllium. "Why only nuclei?" one might ask. Because ***recombination***, the capturing of electrons by nuclei to make completed atoms, did not take place for another 3000 years! Everything was still ***way*** too hot. Also, as already stated, it took atomic nuclei beyond lithium, and maybe beryllium, another 300 ***million*** years to show up.

But why should there be so **very** little beryllium you might ask? As you stack alpha-particles *(don't worry you will soon be doing just that)* to assemble Be-8, note that it is the Be-9 isotope not the Be-8 isotope that dominates in the periodic table. And just why is that? We have the answer. **Be-8 is way, way more than just a little bit unstable.**

The following videos may help one to understand what happens next and how it's done in the assembling of Our Universe **BY** its atoms. First, an advanced look at atomic structure and the way atom's assemble things once the atomic nuclei have captured their electrons in the above mentioned *recombination* process.

YouTube "*Hybridization Theory", (English) PassChem / Sponholtz Productions (31:33).*

Then let's get down to the details of the Periodic Table.

YouTube *"Introduction to the Atom", English / PassChem / Sponholtz Productions* (35:36).

YouTube *"Periodic Table Explained: Introduction", AtomicSchool (14:40).*

YouTube "The Periodic Table – Classification of Elements / Chemistry",

***Khan Academy* (8:55).**

And just in case you missed the following video that I presented in the Preface, this video, and others in the “Stated Clearly” series, will be great support for much of what follows in this book.

YouTube “*What is an Atom and How Do We Know?, Stated Clearly*” (12:14).

And here is a cool way that I found to **just-listen** or **view- listen** to a **mix of videos** like the above. I run while I just-listen and I use my cell-phone with a Bluetooth headset to do it. A *just-listen has the following benefits; One can get a more complete understanding when one finally decides to sit down and view-listen. There is no wasted brain-activity as one works-out and just-listens. There is no wasted time sitting down to view-listen when all one wants or needs is a just-listen.*

1. Open youtube.com.

2. Use the YouTube hourglass-search-feature to find the videos you want from titles or subjects.

3. Create your own personal-mix by downloading each desired video into your automatically-provided, YouTube library. A download-arrow appears under each video on the far right. However, I’ve found that this feature only shows up when my phone is held in its vertical position. The videos will automatically be added to your YouTube library.

4. In order to access and use your downloaded YouTube Library, turnoff Cell-Data and Wi-Fi in Settings.

5. Click on your YouTube app and it will instruct you to use your downloads because your Cell-Data and Wi-Fi have been turned off.

6. Either sit down to watch your mix or, better yet, go outside and just-listen as you walk and or run.

Just-listening is interesting because your brain will imagine what is going on in the video. Also you will be motivated and better informed when you actually watch the thing. PS, *You will get some much needed exercise if you run or walk while you listen. Just saying. If you’ve not done such a thing, I bet you will like learning while you ambulate.*

A Kit and Directions for Assembling Ping-Pong Ball Atomic Nuclei

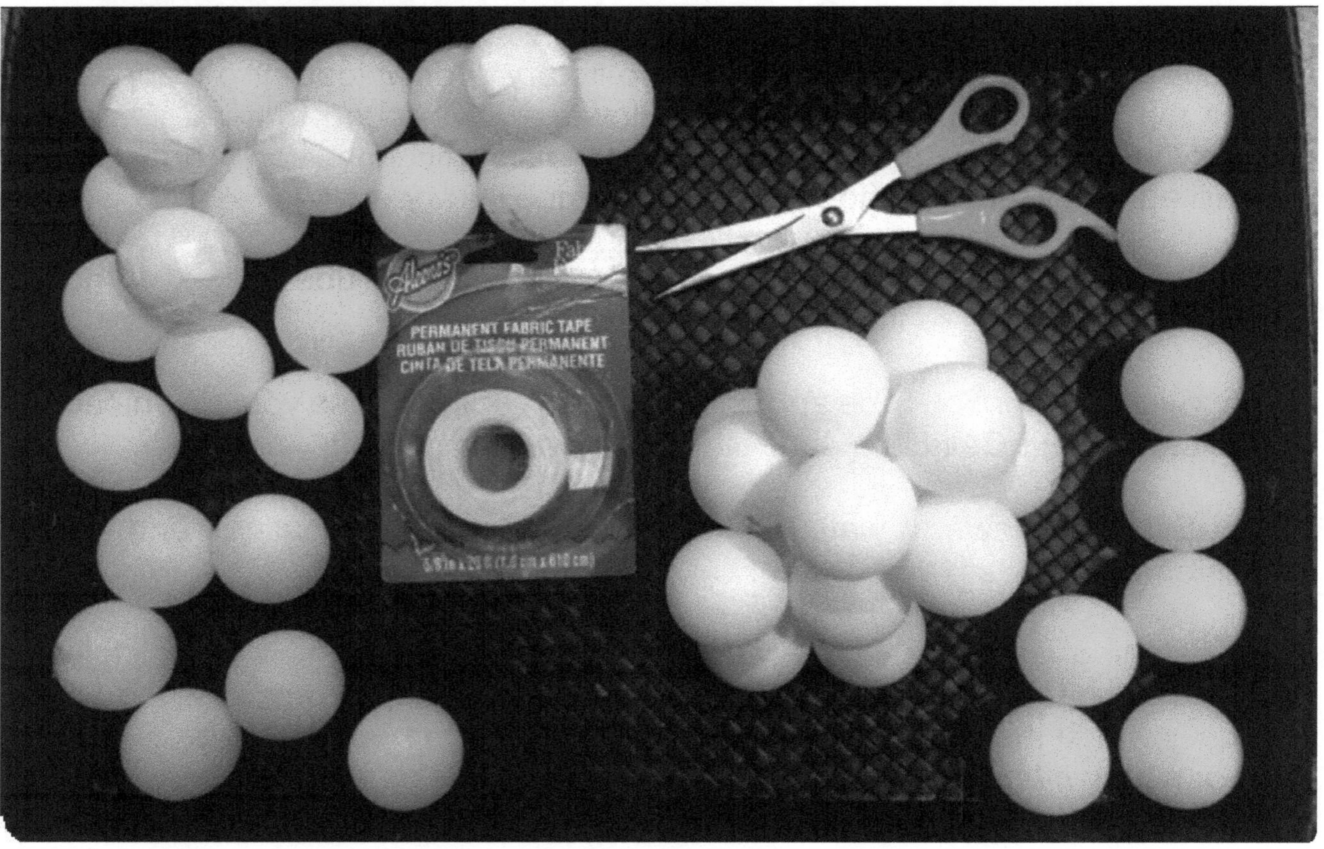

Cut the fabric tape as shown to make three-point placements on 1.5-inch diameter ping-pong balls. Beer-pong balls can work but they are usually less well-crafted and can be indefinite in size. Carefully remove the three protective papers from the tape to leave its glue stuck to the ball, and, then, simply secure the "neutron" or "proton" to an appropriate place on the growing "nucleus". For effective charge separation, I arrange all of the neutrons toward the center of the nucleus with the orange protons arranged on the surface separated from each other.

As an example, the first view of the kit shows the 20 *white* neutrons of the potassium nucleus arranged in semi-spherical form. The second view shows the completed potassium nucleus with its 19 *orange* protons attached, separated and charge-insulated by a central core of neutrons. I unabashedly suggest that this might, in fact, be a way that neutrons and protons could be arranged in atomic nuclei save for the added twist that atomic nuclei must be extremely dynamic things. They are also wave particles, so I am pretty sure there is no way for me or anyone else to create a model that can accurately represent a wave of energy that masquerades as a particle!

This latter concept leads to another one of my hair-brained ideas, *either neutrons and protons do their furious moving about essentially in place within their assembled nuclei, or once fused together by the* ***strong force*** *the constituent* ***quarks*** *of the* ***neutrons*** *and* ***protons*** *freely move about, in a furious near light-speed way, managing charge-separation throughout the nucleus while simultaneously maintaining some kind of mysterious* ***Pauli-Exclusion-Connection*** *to their distantly separated electrons.* See my reviewing account at the end of this treatise for further discussion of this idea. But let there be no doubt, it is almost certain that this idea is both wrong and stupid. Nevertheless, I think there is a lot more that can be revealed as one admires ping-pong balls and the way they generate provocative ***thought experiments***.

For example, a nucleus of protons-only is not possible except for that single proton in a hydrogen atom. *However, unlike protons, neutrons can do the multi-neutron dance*, and they can make it possible for protons to be held together by the strong-force with each proton connecting in some strong-force interacting way to neutrons. The best way I can visualize and think about my ping-pong-ball baryons doing such a thing is for the protons to connect in strong-force interactions with neutrons where each proton on the outer "surface" of the nucleus is connected to proton-separating neutrons. The strong-force is a strong, ***short-distance***-acting force. The electromagnetic force is a strong, ***long-distance***-acting force. Protons in my model nuclei are **connected** to neutrons and ***separated by a short distance*** from the ***repelling-force*** of another proton, just saying. Since one ***really*** can't visualize or properly think about a nucleus, the key words here are "the **best way** I can visualize and think about it." So take what you like and leave the rest. It is a quasi-thought-experiment and I'm certainly no Einstein. But my thought experiment does lead to a hypothesis. The question is, is it a testable one? *Do positive charges in an atom gather at the surface of the atomic nucleus in a Pauli Exclusion dance with an atom's electrons?*

Hydrogen and the Proton

Hydrogen-1, ^1_1H

H-1, stripped of its electron, *is* **the** proton and it is ***identical*** to **the** nucleus of the hydrogen atom. Depending on its role, other designations for hydrogen are H^+, p^+, hydrogen ion, and hydrogen nucleus.

The naked hydrogen nucleus diameter is ~17 femtometers, (fm) or $1.7X10^{-5}$ angstroms (A) and is, as calculated in a figure shown below, roughly 65,000 times smaller than the hydrogen atom itself (diameter ~1.1 angstroms (A), remembering that this particular atom is a nucleus with just one "circling" electron. *The latter, called the hydrogen atom, the protium,* ^1H, *or atomic hydrogen, is the most common* ***isotope*** *of hydrogen. See* **Wikipedia**, ***"Isotope".*** *Two other isotopes of the hydrogen* ***atom*** *are deuterium,* ^2H, *and tritium,* ^3H. *Diatomic hydrogen, i.e., molecular hydrogen, is the most common form of pure hydrogen. It consists of two electromagnetically bonded protons, each with an orbiting electron. See* **PubChem**, ***"Hydrogen".*** *Isotope* is the term used to describe different forms of the same atom. *Molecule* is the term used to describe an atomic structure containing multiple, usually different, atoms. For instance, hydrogen and oxygen gas (H_2 and O_2) are examples of molecules containing the same atom. Whereas, water and carbon dioxide (H_20 and CO_2) are examples of molecules containing different atoms.

Wikipedia, ***"Hydrogen Atom"***.

YouTube "*What is a Molecule?*", Stated Clearly (8:18). BTW, this video is great.

Keeping in mind that atoms and their nuclei actually belong to a realm where everything is unthinkably small, furiously fast and fluid, my ping-pong ball model of an **atomic nucleus**, i.e., one **without** an imaginary electron, is a thinkable,

1.5-inch, static, hard thing. But to represent an entire **atom**, i.e., the **protium**, ***with*** *an* electron *and the* 65,000-times larger electromagnetic forcefield that it generates, my 1.5-inch, static, hard model would need to expand 97,000 inches to a little over 1.5 miles! For vivid comparison, picture a ping-pong ball in the center of a large stadium and then realize ***you are still shorting the far reaches of an impossibly large model of a hydrogen atom by at least a half-mile!***

Appreciating the unimaginable energy relative to the equally unimaginable tininess of the wave-particle we call the hydrogen **nucleus**, and the comparative vast enormity of the electromagnetic forcefield generated by the even smaller wave-particle we call the **electron** are but first steps in grasping the content of our universe, its illusion of hard stuff, and its self-assembly by atoms of everything in it, including us, ***self-aware beings***.

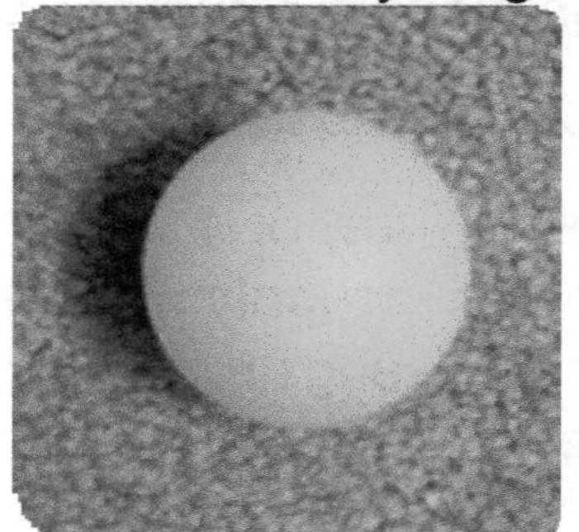

Diatomic Hydrogen

The above model of **molecula**r hydrogen, as will be true for all my atoms represented by ping-pong balls, is shown without its requisite electrons. Why? Once again, it's because for one to accurately represent the ping-pong ball model of the diatomic hydrogen **molecule**, *with its double-electron electromagnetic forcefield,* **each ping-pong ball would need to be in its own stadium and the two stadiums would need to be over a mile apart!** So, sorry, you will just have to imagine that. To differentiate the atom from the diatomic molecule, the two ping-pong balls above are shown separated, but remember it should actually be by a separation of **about a mile.**

Unlike the protons and neutrons of atomic nuclei, which I show to be touching each other, the protons in diatomic molecular hydrogen are not fused. They are separated, yet covalently bonded together, by the vast field of angular force produced and shared by two electrons. Also, it might be good to remember that the electron can be thought of as a *furiously-moving, light-speed, thing* that is simultaneously both a standing wave and a particle, i.e., it is a *waving-particle*.

One estimate that I saw somewhere on the internet has this wave-particle moving, if it is reasonable to think of a standing-wave moving, at more than 10^{31} revolutions per second about its nucleus, or 1 followed by 31 zeros! In other words, electrons are fields of force that act like they are whirling about atoms to create standing waves, or maybe they are better thought of as "particles" that, within their orbital domains, for all practical purposes are, according to the Heisenberg Uncertainty Principle, ***uncertain,*** *and, essentially, everywhere they can be at the same time.*

YouTube *"Heisenberg's Uncertainty Principle Explained", Veritasium"* (4:12).

Molecular diatomic hydrogen (H_2) in the form of pure hydrogen with which we are most familiar, exists as a gas. It consists of two protons that are charge and energy-level balanced by the positive-charge-neutralizing and energy-stabilizing effects of two negatively charged electrons. Without the effect of the two electrons, the positively charged protons would repel each other and would not, as they clearly do, have an attractive need for each other. The attractive need comes about due to the fact that the electromagnetic energy of the hydrogen atom would not be stable without the stabilizing effects of the second hydrogen atom. Stability of the atom in the diatomic state is met by filling the so-called "**K**-shell", i.e., the lowest energy-level of the atom, with two electrons. Filling this first energy level is achieved in diatomic hydrogen by each protium in the diatomic hydrogen molecule making the other one "happy" (see below) by sharing its sole electron with its partner. What does "happy" actually mean? Check it out. It gets complicated with rules and math but, aside from **quantum spin**, at least the rules of Pauli Exclusion are simple enough to understand.

YouTube "*What are the Pauli Exclusion Principle, Aufbau Principle and Hund's Rule", chemistNATE (4:15).*

"The Basic Math that Explains Why Atoms AreArranged Like They Are: Pauli Exclusion Principle", Parth G (10:36).

The fact that hydrogen is "unhappy" with just one electron in its **K**-shell means that protium atoms are relatively few in number. Instead, hydrogen is mostly found in molecular form or in its ionic electron-minus form, the proton, aka, the hydrogen ion, H^+. And, of course, everyone knows H^+ *is* acid. I'll have a lot more to say about this startling fact later.

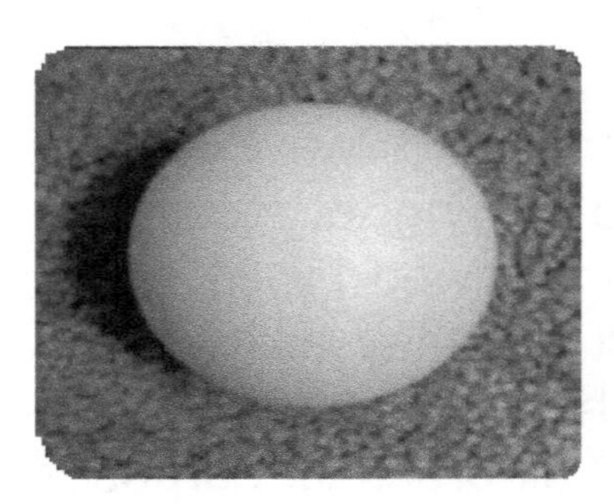

N^0 is the Neutron. It's also designated n or n^0

Neutrons are modified protons. And they are identical to protons except that they are a tiny bit bigger and have a neutral charge. The initial occurrence of neutrons came about when electrons smacked into protons during the first few, super-hot Planck moments of primordial synthesis. Another *surprising*, entirely different way neutrons showed up in our universe occurred much later. And I can't wait to expose this cliffhanger, but we have some other important stuff we need to examine first.

As for the size of the neutron, it's the charge-neutralizing addition of an electron that makes the neutron slightly bigger than the proton (~$1.675X10^{-27}$ vs ~$1.673X10^{-27}$ kg). Neutrons can be either *equal*, *lesser* or *dominant parts* of all atomic nuclei. And with one exception all atomic nuclei depend on neutrons for their existence, the exception being the nucleus of the hydrogen atom itself, which consists of nothing more than a single proton. Only the isotopic forms of the hydrogen **atom** have more, i.e., **deuterium** and **tritium**, with nuclei so special in atom-bomb making that they have been given special names, the **deuteron** and **triton**. The deuteron has one neutron and the triton has two. Or, to say it another way, the hydrogen nucleus has one baryon, the deuteron two and the triton three.

As suggested above the neutrons and protons of *nuclei*, are responsible for most of the mass of *atoms* (one can especially appreciate this relative mass property of atoms when one tries to lift a brick containing $^{197}_{79}Au$ (gold atoms) versus, for example, one of equal size containing $^{27}_{13}Al$ (aluminum atoms). Also, in the formation of atoms, as already noted, I've asserted that neutrons keep the positively charged protons separated or balanced in some other way to be stabilized so that they don't electromagnetically repel each other.

Neutrons contribute to the ***atomic mass***, **mass number** or **atomic weight** but do not contribute to the **atomic number**. For the latter determination, only the number of protons in the atom are counted. And this is the number in ***sub****script preceding the symbols for atoms*, e.g., for gold it's $_{79}$Au. The **atomic number** is also equal to the number of **charge-balancing electrons**. The **mass number**, a number equal to the number of protons plus neutrons in an atom, roughly approximates the atomic weight (*but only if you happen to be on planet Earth, are not under the influence of some other gravitational system and are focused only on the isotope of the atom most closely matching that of the mass number. The* ***atomic mass*** *of the most common isolated isotopic mixture of an atom is usually shown in the periodic table* ***under the symbol of the atom***, e.g. for gold the atomic mass written in atomic mass units, amu, is 196.97. Whereas the mass number is written with the atom's symbol as an even 197 as ^{197}Au or as Au-197. For another example, the atomic number of hydrogen is 1 and its mass number is 1. Whereas its atomic weight or more accurately its actual mass (since weight can vary) is 1.008 Daltons. Daltons were recently revised to "amu" for unified atomic mass units or "Ar", relative amu.

Slideshare.net, "*Relative Atomic Mass*", Siti Alias

Since there is no neutron in hydrogen-1, its mass number, written in superscript as ^{1}H, ends up having the same number as its atomic number, written in subscript as $_{1}$H. The atomic number is the key to understanding the confusion in this setup. The subscripted atomic number NAMES each atom, and always STATES exactly how many positively charged PROTONS and negatively charged ELECTRONS exist in each atom's nucleus. This is clear. However, there is a way to represent this that everyone likes to use because it is much easier and more convenient. But it can get to be confusing.

For example, I can quickly write all of the atoms in the periodic table in this other way to tell me exactly the atom's name, how many protons each atom has in its nucleus, and how many electrons exist in each atom's electron cloud. The following **atomic number** list of the first 10 elements illustrates this convenient method: H-1, He-2, Li-3, Be-4, B-5, C-6, N-7, O-8, F-9, Ne-10., etc., etc., etc. The letter symbol tells me the atom's name. And the number associated with it not only tells me all of the other information to which I just referred, it also gives me some information about where and when each atom was self-assembled by Our Universe! Cool, is it not? And I just whipped that list out from memory, and because it was easy to memorize in this format, I can now continue the list to complete it pretty much in a flash for all 118 of the known elements!

Well, I think this is great, but I find that this format can create a problem in that there is a confusing downside to it. Why? Because, except for H-1, all the rest of the elements in the list represent isotopes that do not and will not ever exist! He-2, Li-3, etc. are like a "horse in a tree", nice to look at but otherwise not very useful! Just like the horse that lacks life, the atom's only come to life when they are complete. Except for H-1 all the viable elements require neutrons for their existence. With neutrons added the **atomic mass-number list** looks more like the following: H-1, He-4, Li-7, Be-9, C-12, N-14, O-16, F-18, Ne-20. Note, I said "more-like". Why? The isotopes can vary. For example, H-2, He-3, Li-8, Be-8, C-14, N-15, etc. also exist. So listing the atoms in the **atomic mass-number way**, instead of in the previous **atomic-number** way, I know for certain the fallowing: 1. the atom's name from its symbol; 2, how many protons are present from its name; 3, how many neutrons are present due to the difference between the total count and the proton count; and 4, how many isotopic forms of the same atom are represented when all of the known isotopes are listed. For example I know, that when I list the atoms in an **atomic number list** carbon, C-6, is number 6 in the sequence and that carbon has 6 protons. I then also know for certain that when I list carbon in an **atomic mass-number list** C-12 is an isotope of carbon that has 6 neutrons.

That's all well and good and I could have been done with my explanation here. Just understanding this much should have cleared everything up. But, oh no, we had to make it even more complicated did we not? Just for hydrogen, *and I mean just for this atom only,* everybody got so excited over the discovery of the neutron and heavy water that they changed the name of hydrogen! They broke the naming rule for this one atom! Add a neutron to any other atom and it keeps its name, but add a neutron to hydrogen, and it is no longer called **hydrogen**! It's called **deuterium** and that's because the new hydrogen atom no longer just has a proton it has a **proton-plus-neutron** that they named a **deuteron!** And then they changed the name of hydrogen again! This time they decided to call it **tritium** because this new form of hydrogen has a **proton-plus-two-neutron** nucleus that they named a **triton**. Thank goodness "they" didn't feel the need to play this name changing game with the rest of the elements!

Oh yes. You guessed it. They are still not done. Let's describe all of this yet another way just for fun. Let's describe it the way those who are really in the know do it. H-1 is the common isotope of the hydrogen nucleus with a single nucleon, a baryon-proton. Whereas, H-2 is a rare hydrogen isotope, called the deuteron with two nucleons, a baryon-proton, plus a baryon-neutron. And H-3 is yet another rare hydrogen isotope called the triton with three nucleons, i.e., a baryon-proton plus two baryon-neutrons. The baryon classification for protons and neutrons in the hadron group is from the **standard model of particle physics**. Fear not. We will straighten this all out before we're done.

Well, I tried. I hope my little nomenclature diversion helped. However, before I end this rant I need to give a shout-out to the **deuteron**. It REALLY **IS** SPECIAL, as you will soon see, and the most special of specialness does not have anything to do with atomic bomb making.

YouTube *"Dalton's Atomic Theory"*, ChemSurvival (4:01).

A common reference to a mass number is heard when it's referred to in carbon-14 dating, where the number 14 is the number of protons plus neutrons in radioactive carbon-14 (^{14}C). The more familiar, nonradioactive, everyday form of carbon is carbon-12 (^{12}C). C-14 has eight neutrons in its nucleus, two more neutrons than are present in C-12. It's C-14's additional two neutrons that give C-14 its unstable radioactive property. This example also illustrates another feature of atoms. Many atoms, including carbon, when isolated are found to contain more than one isotope. The isotopes all have but one atomic number. In the case of carbon that number is 6. But the number of neutrons can vary slightly. In carbon's case, the most common isotope, C-12, contains 6-neutrons, whereas the less common, stable isotope, C-13, contains 7-neutrons and the slightly unstable isotope, C-14, contains 8-neutrons.

YouTube *"Radiometric Dating: Carbon-14 and Uranium-238"*, Professor Dave Explains (6:07).

YouTube *"Nuclear Reactions, Radioactivity, Fission and Fusion"*, Professor Dave Explains (14:11).

Remember, the number of protons in a given atom do not vary. An atom with 6-protons is always carbon, no matter the number of neutrons. The number of protons in an atom can change, but if that happens, it is no longer the same atom! However, with a change in neutron number, the atom's name does not change, but its mass does change, as does its isotope status

Hence, neutron variations in atomic isotopes lead to atomic masses which don't always match that of the most stable isotope because the measured value is often the result of the average weight of multiple isotopes found in naturally occurring isolated samples of elements. Check out a periodic table's atomic masses, which are shown under each symbol in the table, and you will see what I mean. None of the atomic masses (*or weights*) are whole numbers, and some of the decimal numbers diverge greatly from the atom's most common mass number, i.e, that particular isotope's whole-number. *And get this, it really doesn't matter what units one uses here. Actually, you will probably see no units at all for the atomic weights in periodic tables because in modern periodic tables these numbers have all been made relative to 1/12 the mass of carbon-12, i.e., (^{12}C), in its ground state*) P.S, if this has not been confusing, you have not been paying attention.

YouTube *"The Periodic Table – classification of elements / chemistry",* Khan Academy (8:55).

A Review of Basic Atomic Chemistry

Our everyday, nonradioactive, carbon atom has six protons, six neutrons and six electrons. Two of those electrons are in the lowest energy orbitals of the atom, an energy level called the "**K** shell". The **K**-shell of carbon is "filled" with just two electrons. The other four of carbon's six electrons are in orbitals at higher energy levels in its "**L**-shell". But, for energetic stability, carbon has a problem. The **L**-shell is energetically more stable when it is filled by eight electrons which requires the addition of four more electrons. (I like the way ***chem4kids.com*** puts it. **L**-shell is "happier" with its **L**-shell filled to capacity with eight, just as diatomic hydrogen is happier with its **K**-shell filled to capacity with two.) Hence, carbon solves its problem by sharing electrons with other atoms that have similar stability issues, and hydrogen solves its problem the same way. So, one of the many ways carbon and hydrogen solve their mutual instability issues is to grab hold of each other and share electrons. The **K**-shell electrons of four hydrogen atoms share their electrons with carbon's four **L**-shell electrons to give each hydrogen atom its **K**-shell-stabilizing two and carbon its **L**-shell-stabilizing eight.

This sharing process results in the formation of four "covalent bonds", and brings about the self-assembly of the **molecule** called methane, CH_4. In case you don't recognize it, methane is a gas at room temperature, and depending on who was in there last, it's a gas that can often best be appreciated in elevators due to its customary association with smelly $^{32}_{16}S$, sulfur-containing molecules. FYI, methane, itself, is odorless. More important may be the effect of methane on climate. PS, cows are not the problem. Also, check this out, cows can be made to release hydrogen instead of methane if they eat a red, kelp-type of seaweed. Methane IS a big problem but, as I understand it, it's from the melting of our tundra and the huge pockets of it that could one day belch forth from sea beds.

YouTube *"What's the unique Role of Methane in Climate change?", Science Action* (4:24).

So, we see how the building blocks work. With their electromagnetic fields of force intact, two or more atoms, like those in diatomic hydrogen, can *bond* together, at room temperature, to make new *molecules*, but atoms must shed their electrons under intense heat and gravitational pressure to form what is called a **plasma** to allow the nuclei to *fuse* and form new *atomic nuclei*. But, as you will soon see, in a new star the only plasma fusion taking place is hydrogen-1 fusing up in multiple steps to helium-4. That helium fusion process also releases immense amounts of energy, a phenomenon we all know very well from the energy we experience every day from our **star**, the **sun**, as it assembles ***helium nuclei*** from the fusion of **protons**, i.e. **hydrogen nuclei**, and lest we forget from the exploding of **hydrogen** bombs. Just check out the enormous energetic difference that one can witness on **YouTube** of exploding hydrogen fusion bombs versus exploding Hiroshima-type atomic fission bombs .

YouTube *"Hydrogen vs Atomic Bomb", LiveScience (2:03).*

Well, this is all very interesting but I'm a little off topic and getting way ahead of myself. The universe has yet to assemble any *atoms* and has only assembled four atomic *nuclei* before coming to a dead stop, and here I go having it assemble molecules! It's just hard not to get excited when one is "watching" bits and pieces of our universe come together as they self-assemble. Nonetheless, I am actually way *way-* ahead of myself since, in our early universe, neither carbon nor *any other* ***atom*** has yet been completed. And for that electron-capturing process called **recombination** to take place, our universe has to cool down a few billion degrees. And for it to assemble any nuclei, beyond that of beryllium, the existing *nuclei* of hydrogen and deuterium have to do some fancy footwork. Also, stars have to be born, a process that doesn't take place for 300 million years *after* the nuclei of hydrogen, helium, lithium and maybe a tiny amount of beryllium assemble themselves. And, with this last piece of information we are almost ready to reveal the outcome of our **first cliffhanger**.

Protium Deuterium Tritium

Protium atom is the proton plus one electron. Think, "The field of angular force around the orange *'atom'* that I am looking at here is more than a mile larger in diameter than a football stadium!" Protium's nucleus is the proton. **Deuterium atom** is the proton plus a neutron and one electron. Deuterium's nucleus is the deuteron.

Tritium atom is the proton plus two neutrons and one electron. Tritium's nucleus is the triton. (Tritium is a radioactive emitter of weak beta particles. Tritium's beta particles are high speed **electrons**. Beta particles can also be **positrons, i.e., antielectrons, which are positively charged, electron-annihilating antiparticles**

YouTube "*Beta Decay*" Tyler DeWitt (9:56)

Deuterium and the Deuteron

hydrogen-2, $^{2}_{1}H$, also called deuterium-2, $^{2}_{1}D$

It turns out that the deuteron, the special name for the nucleus of the deuterium atom, is a lot more interesting than I first thought. As has already been mentioned, at about three minutes into the first day, our universe approached the goldilocks temperature of about one billion degrees centigrade. It was then that the modified hydrogen nuclei called deuterons began to make things happen. All of the deuterons that could form from the fusion of a proton with a neutron went on to fuse and form the nuclei of $^{4}_{2}He$, helium-4, $^{7}_{3}Li$, lithium-7 and , just maybe, $^{9}_{4}Be$, beryllium-9.

Cliff-Hanger#1 Revealed

In other words, every neutron in our still very tiny universe, i.e., every neutron the electrons could make by slamming themselves into protons, dramatically fused with protons to form deuterons. And, instantly, all of those deuterons disappeared from the scene when they either fell apart because it was too hot or fused with themselves to self-assemble the nuclei of helium, lithium and beryllium. This deuteron disappearing-act, called the *deuteron bottleneck*, led to the *dark ages*, a long era or epoch where nothing else in the way of atomic nuclear assembly happened in our universe until things cooled down a bit. A bit of cooling in this instance was about one billion degrees to a "chilly" 3000.

So, what the heck actually happened at the time of the deuteron bottleneck? It turns out that earlier, when the temperature was much greater than a billion degrees, fusion was raring to go, and it was so raring to go that it would have, in just seconds, fused everything that there was to fuse all the way to the nuclei of iron. And that is what would have happened except for one thing. Helium-4 nuclei (2 neutrons + 2 protons) are necessary intermediates in the assembly of all the larger atomic nuclei. And helium-4 can *only* assemble itself by fusing with two deuterons. Moreover, at temperatures above one billion degrees any deuterons that attempted to form immediately fell apart before they could fuse with another deuteron to assemble helium-4. Also, as the temperature and pressure of our universe decreased, electrons, stopped their primordial slamming into protons to make neutrons. Thus, no more neutrons and no more deuterons meant assembly of our universe had to wait for some changes to be made in the assembly process.,

Three years ago, on the fourth of July in 2012, San Diego had prepared for its traditional, massive, synchronized, quadruple, fireworks display on San Diego Bay. It was supposed to be an hour-long, colorful, state-of- the-art, exploding rocket show. Instead, all four setups went off at once in four, synchronized, ginormous explosions that lit the sky for miles around. Make sure you check out the home videos of this event on YouTube. It's a hoot. You'll be glad you did.

YouTube *"BIG BAY BOOM 2012, MSNBC COC TV SHOW OCT '13"* (4:52).

A San Diego Big Bang

The deuteron bottleneck saved our universe from behaving like the now infamous San Diego fireworks debacle. Instead of a cosmological flash that could have taken everything from deuterons to iron in seconds, our universe and its fireworks show had to wait three minutes until things cooled down enough so that the finicky deuterons could hold themselves together long enough to assemble helium-4. With that done, the fireworks show to end all fireworks shows began in all its, *I wish I could have been there*, glory. But, while it was way better than the fubar in San Diego, it was still a short-lived helium-4, lithium-7 blast that fizzled out within 17 minutes due to a very rapid cooling process brought on by cosmic expansion.

So, that's the deuteron bottleneck and that's how it saved the day but now what? Here we are waiting for the fireworks show to restart, and it doesn't get going again for 300 million years? *Really! Are you kidding me?!!*

Helium

helium-4, He-4

Here again, what one sees in the picture above is my ping-pong ball version of a *nucleus* of an atom. To envision the actual helium atom, one has to think of the enormous electromagnetic field of force that surrounds helium's tiny nucleus, a nucleus that actually consists of an extremely busy set of wave-particles rather than a set of four, incomprehensibly small, solid balls. So, from wherever you are sitting, "see" the actual comparative dimensions of the model atom with its two electrons by decreasing the diameter of the perimeter of the above model's two- electron cloud by about 2/3 of a mile. *What!! Did you just say decrease the size of the helium atom? Why?? That doesn't make any sense! I can see from your ping-pong ball model that the helium nucleus is obviously twice as big!!* True enough, I say, but helium has two powerful protons with two powerful positive charges that pull on helium's two electrons to suck them in closer to the atom's nucleus. It's the electron cloud and the effect of the positive nuclear magnetic force, not the actual dimensions of atomic nuclei that governs the size of an atom. So please, keep in mind the size of the **electron cloud** and the effect that the **nucleus** has on the size of that cloud as you assemble your **atomic nuclei** with ping-pong balls and imagine the size of the atom that your model represents.

YouTube *"Which atom is smaller? (in the entire periodic table) HELIUM"*, chemistNATE (3:15).

Some Cosmological Problems: Dark Matter and too many Deuterons

As already mentioned, our universe could not make the nuclei of helium until it first assembled deuterons. And this assembly is thought to have occurred within 10 seconds to 20 minutes of the first day. Deuterons are thermodynamically unstable and the pressure and temperature at this early moment in primordial nucleosynthesis were just right for most of those deuterons to kinetically **burn** to the **ash** called helium-4. To this day the helium-4 assembled at the beginning of time 13.8 billion years ago is thought to account for most of the helium that currently exists in our universe. It's also thought that most of the deuterium in today's universe came from deuterons that failed to burn in those first primordial minutes. However, as I understand it, there may be a problem with this idea. Based on current estimates of the amount of deuterium in our universe, there shouldn't be so much of it. If most of the deuterons were burned up when two deuterons fused to assemble helium-4, why is there so much free deuterium? That is, how is it that there were so many deuterons left to grab electrons to become deuterium atoms at the time of *recombination*. To recap, recombination is the term used to describe the capture of electrons by atomic nuclei, and the first such event is thought to have occurred about 3000 years after the Big Bang.

YouTube *"All Physics Explained in 15 minutes (worth remembering) Arvin Ash"* (17:14).

Aside from the fact that helium is lighter than air, makes possible, therefore, the amazing "Air Swimmers Shark", and high squeaky voices when helium is inhaled *(You've got to see the wonderful YouTube videos for both of these uses of helium)*, I find the most interesting things to know about helium are its assembly as an ash in stars, its role in alpha particle assembly of new atoms, and its connection to “dark matter” in the deuterium story. Just talk to your phone. I can say, “Hey! Siri!” - - “Air Swimmers Shark” or “Dark matter” or “Helium ash” or “Alpha particle” or “Deuterium”. I still can’t believe how cool this is. Siri, Alexa and Google AI’s didn’t even exist when I first wrote this “masterpiece”.

Regarding dark matter, the universe, as we know it, is assembled from protons and neutrons, but, apparently, once again I am told that there is a problem. Apparently, most of the matter in our universe consists of something other than protons and neutrons. And because no one yet knows what *it* is, *it* has been called *dark matter*. But there is an even more troublesome unknown. It also appears that we have yet to account for the lion’s share of the energy in our universe. Again, since nobody yet knows what that is, it’s been called *dark energy*.

The mass of dark matter has been calculated and the effects of dark energy have been well documented. Check this fact out below in the first video. How was this discovered? Several ways but it was first discovered from a shocker. Planets whirl about stars. The closer to the star, the faster each planet travels as it swings about in its orbit. This effect is all in agreement with what can be calculated from the force of gravity delivered by the star that exists at the center of each planetary system. For example, Earth takes about a year to circle our star, the sun. On the other hand, Neptune, our most distant gas-giant, takes about 165 years to complete an orbit! But now we come to the shocker. Unlike planetary systems, the half trillion stars or so that circle the black hole in the Andromeda Galaxy were recently discovered to travel at about the same rate. Stars in the outer most reaches of galaxies have been found travel just as fast as those near the center! Why? How? We really don’t know but it seems dark forces are the only thing that can account for it! And, as of now, it appears that the dark stuff is still the biggest problem facing cosmologists. The following recently released video reveals in beautiful detail all we currently know in 2022 about The Dark Stuff.

YouTube *“Where Did Dark Matter and Dark Energy Come From?”, History of the Universe* (45:18).

YouTube “Light and Dark 2 of 2 Jim Al-khalili”, Hot Documentry / Jame Hola (1:02:03). I have included a copy of this excellent video, which has, apparently, been abandoned. I use it here only as an example of what can happen with an evergreen AWTbook™. As time passes, new material will become available and old material will disappear. Do Not

Worry. Titles of old videos will bring up ones like it that are more recent. This is exactly what happened when I saw the demise of "Light and Dark Part 2" by Jim Al-Khalili. I don't know what happened. It's kind of sad. I thought it was a great video, but it really does not matter. Another video took its place. "Where did Dark Matter and Dark Energy Come From". However, in this case the video didn't entirely disappear. Jame Hola preserved a copy of "Light and Dark 2". But, I guess, he had to purposefully make it annoying to skirt copyright issues or something. As a result, it has a very distracting background of green rain, and it has such a poor sound quality, the video is even bad for a "just-listening-to". I've retained this video for you here only to show it as an example of what can happen over time to YouTube and other such videos as they are supplanted. As the AWTbook™ technology improves, websites other than YouTube, etc., might be required by law to be dropped or be supplanted because other videos prove to be more useful. Such adjustments are to be expected with our new, rapidly evolving technological world. Fortunately, AWTbooks™ can easily go with the flow.

YouTube *"Neil deGrasse Tyson: What is Dark Matter? What is Dark Energy?", Science Time* (10:04).

A final point about helium, the helium-4 **nucleus** is, itself, the **alpha-particle**, an important form of radiation that can show up in cosmic rays from exploding stars or from the radioactive decay of heavy atoms.

YouTube *"What is Alpha Radiation?", FuseSchool – Global Education* (3:36).

The following nuclei of atoms have been shown to be assembled by a process of sequential alpha-particle bombardment and fusion that starts with **carbon-12**: **oxygen-16**, **neon-20**, **magnesium-24**, **silicon-28**, **sulfur-32**, **argon-36**, **calcium-40**, **titanium-44**, **chromium-48**, **iron- 52**, and **nickel-56**. Note how the mass number of each **nucleus** in the sequence increases by four when an alpha-particle, two protons plus two neutrons, smacks into it and fuses with it.

Suggestion, make a number of helium-4 nuclei with ping-pong balls and then start stacking them, first to form beryllium-8 and then carbon-12, etc., etc. all the way to nickel-56 if you so choose. From carbon-12 that's just the way an old, aging, main-sequence star would do it. This little project will leave you with a cool display.

Note, too, how this process seems to fizzle out when it gets to iron and nickel. When the process gets to iron it is no longer an exothermic thermodynamically unstable process. The necessary energy for iron and nickel assembly in a star, at least with any efficiency, is simply not enough without the star somehow coming up with a burst of additional energy. And small bursts are not enough for a star to fuse up all the other atoms in the periodic table. To say it another way the amount of energy needed is not supported by the alpha-fusion process. So the atoms have to get clever again and do it some other way for us to have a Universe to live in. So, how do stars do it? How do they manage to fuse-up the atoms beyond iron? The answer to this question I leave for you to ponder as another cliffhanger.

But first, let the excitement build. What follows is the answer to **Cliffhanger Two.** Read on.

Cliff-Hanger #2 Revealed

So, we know how helium-4 was assembled by deuterons in the first few minutes of primordial synthesis. And we know, that most, if not all of the neutrons made by the bombardment of protons by electrons in the first Planck moments of our universe were used up making deuterons and the nuclei of helium, lithium and beryllium. And yet we know that all the atomic nuclei beyond beryllium also contain neutrons and the heavier atoms have even more neutrons than they have protons. So, the big question is, from whence do all these new, ***non-primordial*** neutrons come?

Did I hear someone say ***proton-proton fusion***? Yes, you heard that right! I didn't believe it either, protons *can* fuse with protons! *But*, because of electromagnetic repulsion, proton proton fusion is a *very slow* process. Also, protons couldn't fuse until the temperature of the universe had cooled to a point where the energy required to fuse them was great enough to overcome the protons' mutual electromagnetic repulsion, and yet not so great as to disassemble the fusion products before they moved on to the next step, i.e., the fusion-assembly of helium-4. And, most importantly to us, this is where **main-sequence**, helium-assembling stars, such as our sun, play their important, universe-assembling, life creating role.

YouTube *"What Makes You ALIVE? Is Life Even REAL?!", The Science Asylum* (14:07*).*

How so, you ask? Because proton-proton fusion is so slow, main sequence stars, such as our sun, can burn for billions of years before they run out of protons for making helium. If it weren't for this timed-release property of the proton fusion process, our sun would have burned out billions of years ago, Earth and the beings on it would not exist, and our starry sky would, by now, be a dark, lifeless place filled with dead stars and galaxies. Interesting how things all fit together, isn't it?

How the Universe has and Continues to Assemble Itself

A Summary of Details from **Wikipedia.com *"Stellar Nucleosynthesis, an Astronomy WikiProject".***

Let's review. 13.8 billion years ago, or so, a force, apparently coming from nothing or nearly nothing (one Planck length), divided itself into four vectored-energies; the weak, strong, electromagnetic and gravitational forces of The Standard Model. The weak force is the force behind the phenomenon we call radiation. The strong force is the "gluon" wave-particle that "glues" atoms and other things together and makes wave particles such as quarks, protons, and neutrons possible. The electromagnetic force is the force behind the spin and attraction that electrons have for protons. And the gravitational force is the force behind gravity, the far reaching attraction that massive bodies have for each other. Gravity is also the force that continues to have all of us primitive sentient beings stumped.

YouTube *"Visualizing the Planck Length. Why is it the smallest length in the Universe?", Arvin Ash* (9:54).

Jumping ahead many Planck-time units, these same forces assembled quarks, protons, neutrons, electrons, neutrinos, and a host of other wave-particles. A little later, gravity, the force that gave mass to the massless and created immense pressure in our early universe, over the next twenty minutes or so caused the exothermal fusion-assembly of helium and lithium nuclei from protons, neutrons and deuterons. But then electrons stopped smacking protons to make neutrons, protons sucked up all of the available neutrons to assemble deuterons, and all the fusion activity came to a sudden halt for about 300 million years.

During this time Our Universe continued to rapidly expand and in about 3000 years that expansion helped to get things to cool off enough for the nuclei of hydrogen, helium and lithium to harvest their enormous electromagnetic fields of force to become genuine atoms. And then, after a really long waiting period called the **dark ages**, as told by the **Astronomy WikiProject**, at about 300 million years after the birth of our universe, ***stellar-nucleosynthesis*** began.

Proton-Proton Chain Reaction.

In the above picture, going counter clockwise, the bottom two "protons" overcome their electrostatic repulsion long enough to come into contact, and fuse to form helium-2 (the two connected orange balls). Helium-2 is an unstable di-proton that almost **always** decays to reform two separated protons. However, on **rare** occasions, a di-proton will throw off a **positron-antiparticle** and a **neutrino**, thereby (as shown in the next step) transforming itself into a deuteron (the orange ball connected to a white ball). The deuteron, in turn, fuses with an incoming third proton, throws off a gamma ray to assemble itself into helium-3. A second, incoming, helium-3 fuses with the first one to assemble helium-4 plus two free protons. The two free protons then make themselves available to repeat the chain-reaction.

The notations that summarize the proton-proton chain reaction are as follow. A variation involving the intermediate formation of beryllium-7 and lithium-7 also takes place but at a much lower rate.

proton + proton ~~> helium-2 (di-proton) + a gamma ray

helium-2 ~~ > a deuteron + a positron and a neutrino

deuteron + proton ~~> helium-3 + a gamma ray

two helium-3 nuclei ~~> helium-4 + two protons

YouTube *"Steller Nucleosynthesis Explained in 4 minutes",* Greg Salazar (4:27).

As already mentioned, due to electromagnetic repulsion of protons, statistically, proton-proton fusion almost never happens, but it happens enough to keep a main sequence star like our sun burning brightly for billions of years. The thing is, there are scads of protons in our star. You will hear others say that our sun is powered by hydrogen but we know that to be misleading, do we not? A plasma is a plasma because it is too hot for atoms. That is, in a plasma there are only naked atomic nuclei, i.e., atoms stripped of electrons. Yes, our sun is filled with scads of hydrogen but because that hydrogen is nothing but scads of nuclei it's better to call the scads protons. So, even if the half-life for a proton fusing with another proton to become a neutron is a billion years (which it is), that incredibly slow process of "burning" protons is exactly what is needed to keep our sun happy and glowing, pretty much as it is now, for another 6 billion years or so. And, of course, that means we can probably count on our planet being around for that length of time too. Some exciting bumps and grinds will certainly occur during that time, but I guess we will get to miss out on that unless we figure out a way to live forever. And lest you think living forever is such a ridiculous idea, check out the RNAi and CRISPR advances that have already been made in that direction!

YouTube *"RNA interference (RNAi)",* Nature Video (5:07).

YouTube *"Genome Editing with CRISPR-Cas9",* McGovern Institute (4:13).

What with living forever and robots smarter than humans, we old folks won't need young folks. For that matter, as already predicted in ***The Terminator*** **movie**, our robots won't need us humans. On second thought, I think I'll sit this one out.

YouTube "*The Most Incredible Recent fully functioning Female Humanoid Robots 2021",* Unexplained Mysteries (15:03).

Lithium

After helium, the next atomic nucleus to be assembled in the primordial universe is lithium-7.

lithium-7, ${}^{7}_{3}$Li

Note how the four white "neutrons" *insulate* the three orange "protons". Also note how well fabric-tape works to fuse these model baryons. It is thought that most of the present day lithium nuclei were assembled along with helium in the early few minutes of primordial nucleosynthesis.

Lithium, ${}^{7}_{3}$Li, is important in that it is the first *alkali metal* in the periodic table followed in series by sodium, ${}^{23}_{11}$Na; potassium, ${}^{39}_{19}$K; rubidium, ${}^{87}_{37}$Rb; cesium, ${}^{133}_{55}$Cs; francium, ${}^{223}_{87}$Fr.

Lithium's effectiveness in treating mania was first noted by Soranus in 200 AD when he found the town's water responsible for calming people. The element has many other uses including its role in batteries, explosives, glass manufacturing, concrete and the Tesla car.

Although we see in the above model the nucleus of lithium, it's the three electrons and the enormous electromagnetic forcefield that the electrons generate which gives lithium its properties. The three protons are charge-balanced by the three electrons. Two of those electrons are in the lowest energy level, the ***K****-shell level*, or energy level *1*. The third electron is all alone in the next higher energy level, the ***L****-shell level* or energy level *2*. And this is where things get interesting. It turns out that hydrogen, lithium, sodium, potassium, rubidium, cesium and francium all end up with the same electron schematic problem where a single, unstable electron hangs out in the outermost, most highly energetic shell and orbital.

So what? So what is, all of these atoms are charge-balanced by electrons that leave a single, unstable electron "whipping, spinning, waving or perturbing empty space as a wave/particle" in its electromagnetic cloud by itself at the highest energy level, and it's this unstable electron that is the problem. So, just as hydrogen does, these atoms either shed that single, unstable electron to water to become ionized Li+, Na+, K+, Rb+, Cs+ or Fr+, or they stabilize it by assembling themselves into ionically bonded molecules, just as sodium does in table salt. Are you beginning to see how atoms assemble molecules? But hydrogen is unique. Unlike the alkali metals, hydrogen can also form true covalent bonds, just as it does in methane, CH4. And in its ionic form, H+, it takes on a very special role, and that is, it ***is*** the H in pH. More on that later.

Hydrogen and the Alkali Metals

hydrogen-1, $^{1}_{1}H$ (Its nucleus is the proton, the primary "brick" needed to build a universe, H+ ion, acid, the H in pH, HCl, CH4, $1s^1$)

lithium-7, $^{7}_{3}Li$ (Li+ ion, LiCl salt, batteries, concrete, psychoactive drugs, $1s^2 2s^1$)

sodium-23, $^{23}_{11}Na$ (Na+ ion, NaCl table salt, sodium hydroxide basic solutions, saline solutions, $1s^2 2s^2 2p^6 3s^1$)

potassium-39, $^{39}_{19}K$ (K+ ion, KCl salt, blood pressure ion-balance with sodium) $1s^22s^22p^63s^23p^64s^1$

rubidium-87, $^{87}_{37}Rb$ (Rb+ ion, RbCl salt, thin film batteries, $1s^22s^22p^63s^23p^63d^{10}4s^24p^65s^1$)

While I was assembling rubidium-87 using my own rules, rather than the current version of quantum mechanics, I was struck by the idea that maybe, protons and neutrons had come up with rules for a simple assembly format similar to mine. What you see on the right in the picture above are the needed 50 white "neutrons in rubidium-87. My charge separation rule leaves the 37 orange "protons" of my "rubidium nucleus" on the outside of a core of its 50 white neutrons. It is practically certain to be the case that atoms structured with neutron cores is a stupid idea, which probably accounts for the fact that I've not seen this arrangement suggested elsewhere. (At least not exactly. Checkout ***"Halo Nucleus",*** **Wikipedia.org**)

Nevertheless, having the protons on the outside of a charge-separating core of neutrons has great appeal for assembling the ping-pong ball version of atoms with fabric tape and for relating to the uninitiated the key role that neutrons play in atoms. My configuration would also, it seems to me, leave the protons in the optimum position to interact with their charge-balancing electrons. Moreover, if nothing else, my configuration makes it easy to create ping-pong ball models of atomic nuclei - models I think bring a new appreciation and comparative perspective to the creative genius of the protons and neutrons of the atoms that are, even now, at this very minute, self-assembling our universe. Okay, so it's wrong but it helps one think, I think. ***Physicsforums.com*** has a discussion about this. It appears nuclei could have a shell structure to match an atom's electrons but it is complicated and apparently leads to nuclei that are definitely not spherical. The uranium, U-238 nucleus has been mapped using the orbiting proton of the neptunium, Np-239 nucleus, which suggest the U-238 nucleus maybe cigar-shaped. In other words don't hold your breath or worry about it for now. My models are just that, models. And they have helped me to think about atoms. I hope they can do the same for you.

cesium-133 ${}^{133}_{55}Cs$ (Cs+ ion, CsF salt, Cs-133 is the only stable isotope amongst a total of 39. Cesium-137 and strontium-90 are the major radioactive Chernobyl-disaster isotopes.

Cs-133 is used in atomic clocks. Due to comparative instability, metallic Cs-133 produces a more powerful reaction in water than rubidium or the other less massive alkali metals, see YouTube.com videos. $1s^2 2s^2 2p^6 3s^2 3p^6 3d^{10} 4s^2 4p^6 4d^{10} 5s^2 5p^6 6s^1$).

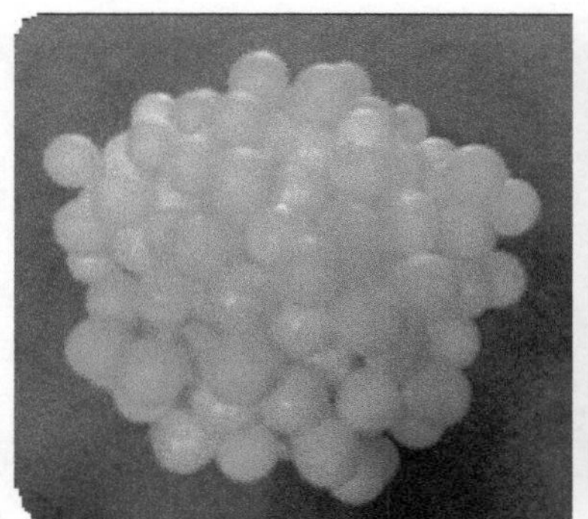

Francium-223 ${}^{223}_{87}Fr$ (Fr+ ion, Fr-223, the only "stable" isotope, has a half-life of only 20 minutes. It occurs naturally in uranium deposits as the result of alpha decay of actinium-227. It's estimated that there are only 30 grams of francium in the earth's crust at any one time. It is the last naturally occurring atom to be discovered. $1s^2 2s^2 2p^6 3s^2 3p^6 3d^{10} 4s^2 4p^6 4d^{10} 4f^{14} 5s^2 5p^6 5d^{10} 6s^2 6p^6 7s^1$).

Comparing the 1.5-in ping-pong-ball nuclei of H-1, Li-7, Na-23, K-39, Rb-85, Cs-140, and Fr-223, and the $1.7X10^{-5}$ Å quantum measurement of the real H-1 nucleus to its 1.1 Å atom as re-expressed in miles.

H Li Na K Rb Cs Fr

1.5 inch (40 mm) ping-pong-ball-scale ***nuclei***

1.5 inch (40 mm) ping-pong-ball-scale H-1 ***atom***

1.1 Å / $1.7X10^{-5}$ Å = 64,706 X the atom is > the nucleus!

64,706 X 1.5 in = 97,058 in/12 in/ft = 8,088 ft

8,088 ft / 5,280 ft/mile = 1.53 miles

So, an H-1 ***atom*** with a 1.5 inch ping-pong-ball nucleus would swell to 1.53 miles to accomidate its one "electron"!!

The francium-223 neutron core and its completed nucleus

Francium is unstable. I found it funny that my model of the francium nucleus had already begun to melt by the time I took its picture.

boron-11, $^{11}_{5}B$, $1s^2 2s^2 2p^1$ (Possible, minor, primordial assembly. Otherwise, this somewhat scarce element is made by nuclear fission spallation, cosmic rays blasting larger atoms). Its primary elemental use is in fibers to make high-strength materials.

Timeline of Self-Assembly

As has already been mentioned, most of the extant hydrogen and helium nuclei self-assembled in our universe during its first few seconds and minutes. Some lithium, and some think a little beryllium and traces of boron nuclei also assembled at this time but these nuclei and their atoms also continue to this day to show-up occasionally when cosmic-rays break up preexisting atoms of carbon, nitrogen and oxygen. Carbon, and atoms of larger atomic number up to and including iron, also continue to assemble themselves in stars.

Cliffhanger revealed

As for other atoms, most of the ones larger than carbon began self-assembling much later in larger stars in the creation time-line and beyond iron only by way of ongoing processes that occur with the help of the enormous energy produced in supernova and other cosmic cataclysms such as those created by the colliding of neutron stars. As already discussed, it takes a lot of energy to stick protons together, and as atomic nuclei get larger so, too, does the energy required to get them to fuse. Until the assembly process reaches iron, fusion is actually exothermal, meaning the fusion process yields energy. After iron, things start to go endothermal, meaning it takes the addition of energy to get the fusion process to take place. And the bigger the nuclei get, the larger the energy that is required.

As already noted, main sequence stars, such as our sun, burn hydrogen to helium-4 ash. Billions of years in the future when our star runs out of helium, it will cool down and expand itself into a red giant. Such stars have the gravity and temperature sufficient to assemble some new atomic nuclei by first "burning" two helium-4 nuclei to produce beryllium-8 ash and through a second process, fusing an additional helium-4 with beryllium-8 to produce carbon-12. Both of these fusions lead to the release of the large amounts of energy that keep red giants alive. But, no doubt, having consumed a number of circling planets as they increase their size to enormity, red giants get their just deserts as they don't live nearly as long as the main sequence star that gave them birth. However, before they die or, actually, before they fizzle into brown dwarfs, red giants also burn carbon-ash nuclei by fusing them with helium-4 to produce oxygen and more heat energy. We should thank our lucky stars for that last part!

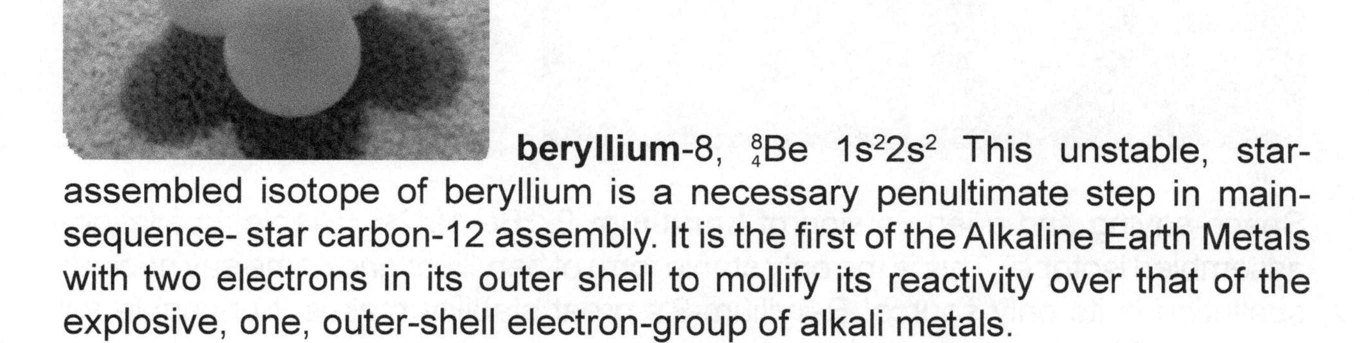

beryllium-8, $^{8}_{4}Be$ $1s^22s^2$ This unstable, star-assembled isotope of beryllium is a necessary penultimate step in main-sequence- star carbon-12 assembly. It is the first of the Alkaline Earth Metals with two electrons in its outer shell to mollify its reactivity over that of the explosive, one, outer-shell electron-group of alkali metals.

Beryllium-8 has a very short half-life but it is a necessary intermediate in the assembly of carbon, $^{12}_{6}C$, in main sequence stars such our sun. Beryllium-9, $^{9}_{4}Be$, the only stable isotope of beryllium, was made primordially. And, because of its stability, it can't be used for the assembly of carbon. This fact, and the absence of primordial beryllium-8, accounts for the absence of primordial carbon, $^{12}_{6}C$ nuclei. Interesting, isn't it?

$2\,^{4}_{2}He \rightarrow ^{8}_{4}Be + ^{4}_{2}He \rightarrow ^{12}_{6}C$ carbon-12, $^{12}_{6}C$ $1s^22s^22p^2$

Space-saving and open version of **beryllium**-9, ${}^{9}_{4}Be$ $1s^2 2s^2$ (Stable, spallation-assembled isotope). This is the only stable form of beryllium and some say nuclear spallation is its only source! Beryllium-9's great stability makes it unusable for the assembly of carbon-12.

Astronomicalreturns.com *"How the Universe Synthesized JWST's beryllium".*

YouTube *"Beryllium Unearthed"*, davidvblack (8:30).

The following video was just posted as I edited this work on 3/10/2022. It has some revelations pertinent to the early, Universe, self-assembly process and the roles of Li-8, Be-9 and B-10 in the sequence of that assembly:

YouTube *"Was Our Current universe Already Inevitable At one Second Old?"*, History of the Universe (45:37).

A Review of my off-the-wall ideas concerning neutron cores and the distribution of quarks in atomic nuclei.

I think the following may be a new idea. But the reader should know it is only a new idea because it is very likely completely wrong. *Once fused, the "up and down" quarks (the wave particles that initially assemble and constitute the inner workings of protons and neutrons) might be free to move about within the entire nucleus of an atom. In other words, if neutrons act as insulating proton-charge-separators, does that mean (once a nucleus is assembled, and protons are fused with neutrons by the strong force) that the charge distribution remains as is, or do the positive charges of the nucleus move furiously about at the surface of a proton core much as the negative charges in the electrons do in the atom's electromagnetic force-field? In such a model, the nucleus would be a blur of activity, especially at the surface, where protons and neutrons only hold their identity for as long as the light-speed motion of quarks allow. The balance of positive and negative charges would remain constant but the things we call protons would, in a process approaching light-speed, flash on and off at the surface of the nucleus with the quarks of neutrons maintaining their spin neutrality and role as insulators in an inner core.*

I have no idea if this model of the "neutron core" has any connection to reality nor do I know if it's a testable hypothesis that would even be interesting to anyone but me. But my model does illustrate the role of neutrons in moderating the repulsive force of protons that would otherwise make it impossible for them to take part in the formation of a nucleus. With protons on the outside of a neutron core might they not obey the same Pauli Exclusion laws relegated to the electrons in their charge balancing act that brings a state of overall charge neutrality to a nucleus? In other words, this might mean that an electron-cloud surrounding a nucleus might be complemented by a charge balancing proton-cloud that's moderated by neutrons so that the protons can match the much larger electron cloud's Pauli Exclusion dance. I'm so sorry. This is surely nonsense proffered by an idiot, but I couldn't resist my imagination exposing my ignorance. Also, if any of what I brought up here is true, just what does a nucleus look like when it is stripped of its electrons in a plasma?

Some reflections on the fact that the proton, also known as the hydrogen ion or H^+, is acid, whereas the potassium ion, K^+, and the other alkali metal ions are just that, ions

Everyone knows about acid but not everyone knows what it is. Most know how acidic things like vinegar and lemons taste sour, and they know that acid has something to do with pH. But what exactly is pH? It's *just* the negative log of the hydrogen ion concentration. "Yeah, right" you say, "speak English".

If hydrogen is not covalently bonded to itself or to another atom, it's got a problem. It is unstable and has to get rid of its one electron, somehow. In water solutions, hydrogen hands off its electron to H_20 to help balance the partial positive charge water gets from its two covalently bonded hydrogen atoms. But that leaves H^+ in solution and the more H^+ there is, the more "acidic" a water solution is said to be.

The concentration of H^+ in water can be measured with a pH meter. For example, say the concentration of H+ is 0.001 Molar, i.e., 10^{-3} Molar. If we just drop out the number 10 in the notation 10^{-3}, refer only to the -3 superscript and multiply the superscript -3 by -1 to make it a +3 rather than a -3, we have a 3! Bingo, the pH is 3! And *that's the negative log of the hydrogen ion concentration*. But pH is a little tricky. One has to remember that the *higher* the pH the *less* the solution is acidic! For example, a pH of 2 is ten times *more* acidic than a pH of 3, and a pH of 4 is ten times *less* acidic. I'm guessing some of my readers may have not bumped into moles recently, neither the underground variety or the mole as used in chemistry. If not, before going on check out the following cool video that will clarify it for you, *forever*.

YouTube "*Avogadro's Number, The Mole, Grams, Atoms, Molar Mass Calculations – Introduction", The Organic Chemistry Tutor (17:58).*

Why is pH important? It is one of the most important properties of the proton, the nucleus of the first atom to be self-assembled and primary building block of our universe. Nothing else of importance to us would have happened if it hadn't been for the acidic property of the hydrogen atom when it hands off its electron to water! Of course, *absolutely* NOTHING would have happened without the nucleus of hydrogen being *the* primary building block of our universe.

So H^+ *is* acid and it *is* the H in pH. But what is acid, actually? We all, sort of, know what acid does. We all know about hydrochloric acid, sulfuric acid, nitric acid, acetic acid, etc. And probably many reading this know it's the H in HCl, H_2SO_4, H_2NO_3, and CH_3COOH that does the damage. But how? When one puts a strong acid, such as hydrochloric acid, HCl, in water almost all of the HCl ionizes, meaning most of the H in HCl sheds its electron to water to become H^+ and Cl^-. If the concentration of HCl is 0.1 Molar, the concentration of H^+ will be 0.1 (10^{-1}) Molar as well. Taking the negative log of 10^{-1} equals 1 or, in other words, adding that much HCl to water gives us an acidic solution with a pH of 1. That's about as acidic as the stomach-acid we all use to digest food. So, what we are actually talking about when we talk about acid is the ability of hydrogen atoms to work with water and enzymes to break molecular covalent bonds by a mechanism called hydrolysis. Long story short, lithium chloride (LiCl) and the other alkali metal salts, including NaCl table salt, can't do what HCl does. They can't hydrolyze things.

Okay, you've got me. I've been caught. "What about Lewis Acids?", you say. I say, "I don't know. You tell me".

YouTube "*What's a Lewis Acid and Lewis Base?", chemist NATE (4:27).*

But, let's go back to that single electron in the outermost energy level of the alkali metals. What can these metals do with that single electron? Take lithium, ^7Li, for example. The distribution of the three electrons in lithium can be written $1s^22s^1$. That's just shorthand, in little superscript numbers, for two electrons in the s-orbital of lithium's **K**-shell, and for the one, lonely electron in the s-orbital of lithium's **L**-shell. For clarity, let's take another example. Let's do sodium, ^{23}Na. In the chemist's shorthand that's $1s^22s^22p^63s^1$ or two electrons in the s-orbital of sodium's **K**-shell (designated $1s^2$), two electrons in the s-orbital of its **L**-shell (designated $2s^2$), six electrons in the p-orbital of its **L**-shell (designated $2p^6$) and one electron in the s-orbital of its M-shell (designated $3s^1$). Go ahead and count them. You will see that the little superscript numbers add up to the 11 in the subscript of sodium-23, ^{23}Na.

To repeat, the important thing for this discussion is the lonely electron in the outermost shell of the hydrogen atom and in the group of atoms called the alkali metals. Think about it. These atoms all love to form ionic bounds with chlorine, $^{35}_{17}Cl$ to form HCl, NaCl, KCl, RbCl, CsCl, and FrCl. But why? And why is only one of the chlorine molecules, HCl, an acid? The other atoms are called alkali metal atoms and their chlorine molecules are all salts, of which sodium chloride is an example.

The answer to the first question is the solution to the single electron problem. **To make chlorine**, $^{35}_{17}Cl$, "happy" and in line with the Pauli Exclusion principle, hydrogen and the alkali metals share their single outer-shell electron with chlorine to give chlorine's seven electrons in its outer-shell the stable number of eight electrons. The shorthand method for showing electron distribution illustrates this outcome for chorine: $1s^2 2s^2 2p^6 3s^2 3p^5$. The p-orbital in chlorine's number-3 shell (also called **M**) needs an electron for stability. The electron that chlorine gets by sharing in an ionic covalent bond with hydrogen, or any of the alkali metals, gives chlorine the stable number of eight electrons in its outer-shell (two electrons in 3s and six electrons in 3p) and relieves hydrogen and the alkali metals of their lonely, unstable single electrons in their outer most orbitals. The net result of this ionic bonding is the complete ionization of hydrogen and the alkali metals to H^+, Na^+, K^+, Rb^+, Cs^+ and Fr^+ when their ionically-bonded molecular forms, HCl, NaCl, KCl, RbCl, CsCl and FrCl, get dissolved in water. And, since chlorine has taken on an electron to stabilize itself, it also finds itself in an ionized, negatively charged form, Cl^-. This, now, gets to the second question. Why is HCl an acid whereas NaCl, KCl, RbCl, CsCl and FrCl aren't?

Water, H_2O, is arguably the most magical of Earth's many molecules. And, obviously, water is nothing like Na_2O, or K_2O, etc. Without getting even more technical, adding more H^+ to water creates a condition where the original baryon building-block of the universe, H^+, can play its unique role as the proton. Ionized, free of its electron, H^+ can behave uniquely to react with other atoms to make nonionic, covalently bonded, new molecules, or behave like an alkali metal to make ionically bonded ones. To say it another way, alkali metals are atoms that have themselves been assembled by H^+ (i.e., protons) and unlike hydrogen can only make ionic bonds with other atoms. I know, that can be a little confusing but the following should help clear things up.

So the H^+ ion is very interesting but that's not to say that the alkali metal ions are not also plenty interesting! For example, add elemental sodium to water and what do you get? You get a tremendous explosive reaction with the release of a lot of heat, energy and a sodium hydroxide basic solution that is the opposite of an acidic solution. That's what you get!

In water, sodium is extremely unstable. It instantly and almost completely ionizes to form the alkaline base, Na^+OH^-. Why? It's that single electron problem expressing itself in a violent reaction. Sodium sheds its single outer shell electron to water and reacts with H_20 molecules in violent, heat-generating reactions that produce Na^+OH^-. But what happens to water's other hydrogen atom you might ask? One of the hydrogen atoms ends up in OH^-, whereas the other hydrogen atom combines with a second hydrogen atom to blow off as H_2, molecular hydrogen gas. That's interesting, but you ain't seen nothin' yet! The high heat produced by the exothermic reaction that occurs when elemental sodium meets water causes the hydrogen gas to ignite! When things get hot enough (check out Ray Bradbury's ***Fahrenheit 451***), H_2 gas molecules will start jiggling and bouncing around enough to react violently with atmospheric O_2 molecules to produce fire-filled, forceful explosions.

Don't believe me? Well, I have a treat for you! Take a look at the **YouTube** video of the alkali metals as they react with water. This is *so* wonderful that if you have, previously, never enjoyed chemistry, I think you might get interested after you see what happens when fabricated metallic forms of the atoms Li, Na, K, Rb, Cs, and Fr are tossed into water! But what happens when plain H_2 gas is mixed with water? Practically nothing, it just bubbles. And why is that? Unlike the alkali metals, H_2, with just the two electrons it needs in its K-shell to be completely stable, is kinetically stable as long as things are not too hot and molecular oxygen, O_2, is absent. However, get H_2 and O_2 together, light a match and watch out!

YouTube "*Reaction of Alkali Metals with Water*", Fathima Yusuf (3:15)

Really, be careful!! All of these explosions are extremely dangerous and have caused serious injury and death. For an example of the extreme hazard with respect to the reaction of hydrogen with oxygen, have a look at the YouTube video of 1938 Germany's Graf Zeppelin called the Hindenburg, and its disastrous demise. It is truly amazing but it's not for the faint of heart.

YouTube "*Rare Hindenburg Disaster Footage in Color", LAKEHURST (5:39).*

You might ask, "Why are you spending so much time talking about all this chemical reaction stuff, especially the acid and base properties? How is this information important to the self-assembly of our universe by atoms?" The acidic property of the proton in its hydrogen ion form is a necessary property for the many coupled enzymatic reactions that take place involving the hydrogen ion. The same is true for the OH^- ion. Organic molecules and life itself, at least as far as we know it, could not exist without the acidic and basic properties exhibited by H^+ and OH^-. If it weren't for the innate way electrons behave when paired with the protons of hydrogen and oxygen, I wouldn't be writing this because I wouldn't be here, and neither would you! So, I think there is one last thing regarding the acid and base properties of water about which I should try to be clear.

It turns out that the electrons whirling, wave-standing or doing whatever it is they do about the hydrogen and oxygen nuclei of the water molecule always lose their attraction to each other at a low rate. In other words, ordinary water will always have a little H^+ and OH^- hanging around at a low concentration that just happens to be about 0.00000001 (10^{-7}) Molar. That is, this is approximately the concentration of these ions that will naturally equilibrate in water. Or to say it yet another way, a little bit of the water molecule always naturally falls apart (ionizes) to about 10^{-7} Molar H^+. Remember, pH is the negative log of the hydrogen ion concentration, and that tells us to eliminate the 10 in 10^{-7} and multiply the exponent -7 by -1 to make the -7 a +7, to get pH 7. But decrease the H^+ concentration by five exponents and increase the OH^- by five and one will arrive at the basic pH of 7 + 5 = pH 12. So what happens to a protein when the OH^- concentration outweighs the H^+ concentration by such a significant amount? Just as a high H^+ acid concentration can bring about dissolution of a protein to its constituent amino acids by *acid* hydrolysis, a high concentration of the basic OH^- ion can bring about a similar dissolution by *alkaline* hydrolysis. (See YouTube videos of acid and base catalyzed hydrolysis of amides for a demonstration).

YouTube "*Introduction to Enzymes and Catalysis / Chemical Processes / MCAT", Khan Academy"* (6:05).

Explaining the difference between thermodynamic and kinetic stability

Most of the time when one bubbles hydrogen gas through water nothing but bubbles happen. But lest you forget, while hydrogen in an oxygen atmosphere can be kinetically stable, thermodynamically it is always VERY unstable.

If you don't yet get the difference between kinetic stability and thermodynamic stability, do the following. Get some plastic cups. (I always used broken glass-beakers when I taught this subject to university students, but plastic cups will prove to be much less messy.) Place the cups as close to the edge of a table as possible. As long as there is no energy added to this "system of cups on a table", the cups will just sit there. And they will happily sit there for the rest of your life! And that will be true until you or something else adds some external energy. In other words, the cups in this example are kinetically stable, just as is hydrogen, H_2, gas in an oxygen atmosphere. But add energy to the cups, such as your fists pounding on the table, and watch what happens. If you've done it right, the cups will come flying off the table to go bouncing to the floor. Therefore, although the cups on the edge of a table can be said to be kinetically *stable*, clearly, they are thermodynamically *unstable*. In other words, with gravity's help they have **potential energy** but won't go anywhere until some energy is added. H_2 gas also has the property of being kinetically stable but thermodynamically unstable in earth's oxygen atmosphere. But watch out if you should accumulate enough of it and light a match, or happen to add some energy in the form of a little spark to a hydrogen-gas dirigible!

Some reflections on Energy, Heat, Temperature and Pressure

Just so we are on the same page, I think I should try to answer some questions about *energy, heat, temperature and pressure.* We are all very familiar with the effects of heat energy but I'm guessing most of us don't think a lot about the nature of heat. For example, a physical way to think about heat is to visualize it as the jiggling of atoms. That's probably not what the average person comes to when they think about heat, but to show how this definition makes sense, I'll use ice for an example.

We judge ice to be cold because the atoms in it aren't moving very fast. When the frozen water molecules share the information about their slowed jiggle with our nerve cells, the electronic message we call "cold" is interpreted as such by our central nervous system. The water molecules in the crystalline structure we call ice will even let our finger freeze and join with it. So, don't even think about putting your tongue on a frozen pipe.

YouTube *"A Christmas Story-The Triple Dog Dare Clip (HD)"* MrHDmovieclips (3:40)

However, if other atoms, such as those found in hot air come in contact with ice, the excited, jiggling atoms in the hot air will interact with the sluggish frozen water atoms which will get those frozen atoms to increase their icy rate of movement and jiggle until the ice begins to melt. With still further increase in the rate of movement and jiggle, the watery atoms can begin to boil and even phase-transition into steam.

YouTube *"Phase Diagrams of Water and CO2 Explained – Chemistry – Melting, Boiling and Critical Point"*, The Organic Chemistry Tutor (10:27).

YouTube *"Vapor Pressure and Boiling",* Wayne Breslyn (1:54).

The more atoms in us move about and jiggle, the hotter we are. **Temperature**, not to be confused with **Heat**, is just a comparative numerical way of describing the average effect of heat energy or atomic jiggle delivered by a collection of atoms or molecules to those in a thermometer. As people usually say it, "My Temperature says I'm hot"! I guess that sounds better than saying, "My temperature says my atoms are over jiggling"! But the jiggle reference *would* be more informative. More exactly, in thermodynamics, *Temperature is the measure of the average heat energy that is available to do work.* Think, just one molecule of water has a calculable Heat energy and so one can get more energy with more molecules. For example a cup of coffee obviously contains much less Heat energy than an ocean, but the cup of coffee could have exactly the same Temperature as the ocean. Got it?

YouTube *"Heat and Temperature"* Professor Dave Explains (4:43)

"Okay, Frank", you might ask. "How does heat arrive at Earth from our star, the Sun? Seeing as how there is a 93 million-mile "vacuum" between the Earth and the Sun and no atoms to be found that can jiggle about in outer space to transmit heat, how does the sun's heat get here?" "Good question", I respond. It arrives here from the energy found in waving wave-particles. Radiations in the form of visible light, ultraviolet light, infrared light and various other photonic forms of heat producing waves strike the earth's atmosphere, as well as terra firma and are very own selves, to cause the jiggling of our surrounding and constituent atoms to the point that we call the day a hot one and we can't wait for a cloud, or nice, cool, marine layer to show up.

Just like a thermometer, a pressure-gauge is just a numerical device to measure the force generated by the combined effects of the rate of atomic jiggle and the abundance of confined atoms. Check your tires with a pressure gauge before and after you run your car. The pressure goes up after you run your car because the temperature inside the tire has increased after you run it. Likewise, you can increase your tire pressure simply by adding more air to the tire.

The ideal gas law, P=nRT/V or as it is usually written, PV=nRT, is an easy way to think about pressure. This equation used to baffle me as much as it did my fellow students, but now I see that it is easily understood and is just a way to summarize the above definition of pressure under ideal gas conditions.

YouTube *"Kinetic Molecular Theory and the ideal Gas Laws"*, Professor Dave Explains (4:43)

You Tube *"Boltzmann's Constant / Physics", Khan Academy* (9:45). This video is useful if you want to understand the pressure equation and how to manipulate R to yield $k_{b,\ i.e.,}$ the more precise value of R that is one of physics most valuable constants.

For example, for the sake of argument, let's add to the confusion of who we are, and where we exist, by thinking of the universe as an expanding balloon. It totally isn't that, and it's certainly not an ideal gas, but the gist of the analogy seems to apply adequately enough to get the idea across. So just relax and think in generalities. Yes, I know, Our Universe may be expanding a little like a balloon, but it's a spooky one because it requires one to address the very troubling thought of our "existence" expanding into "nonexistence" among other things.

At any rate, undaunted, I'll use the PV=nRT analogy as my aid to understanding. When our universe started, it was vanishingly small, i.e., the volume "V" notation in the equation P=nRT/V must have been at Planck Scale, a ***vanishingly*** small number indeed. So, all of the energy, and therefore, too, all of the wave-particles as they came into existence in the early moments of our primordial universe, would have been **unimaginably** confined. The vast number of wave-particles that formed (the abundance "n" notation in the PV=nRT equation) would have been furiously trying to break out of confinement. As already discussed above, the furious jiggling in our primordial universe made all the energy in it Planck-Scale hot (the temperature "T" notation). The force exerted by the crazed attempt by all the energy in our universe trying to break out of very tiny "V" confinement would have made the pressure (the "P" notation) literally higher than it would have been if our entire Universe had been trapped in a Coronavirus.

YouTube *"Visualizing the Planck Length. Why is it the Smallest Length in the Universe?", Arvin Ash (9:58).*

Therefore, all the temperature and pressure measurements at the beginning of Our Universe, the large numbers I referred to earlier, would have been large beyond belief due to the unimaginable energy released at the time of the Big Bang. To understand the analogy I'm using with the ideal gas law one is only left to explain the mysterious R in the above equation. R is the constant that one can derive from the slope of the straight line that one gets when one takes a series of PV/nT=R measurements under carefully controlled, standard conditions where

Temperature is in Kelvins, Volume is in Liters, Pressure is in Atmospheres and Avogadro-Moles are in Grams. For example plot P-measurements on the X-axis and corresponding T-measurements on the Y-axis while holding V constant at one-liter and grams equal to one-mole of the gas measured and you should get a straight line with a slope equal to the constant, R. That's it. That is all there is to the ideal gas law or P=NRT/V. In this form it is easy to see from the equation that P, the pressure, is inversely related to V, the volume. That is, put the same number of gas atoms, N, in fixed, smaller volumes, V, and the pressure, P, will get progressively higher. Also, the more the atoms jiggle, i.e., the higher temperature, T, results in a proportionally higher Pressure, P.

To be clear, the Ideal Gas Law equation is good for understanding and acquiring accurate values of Pressure, Volume and Temperature as they occur on Earth today, but it is only good as an analogy for a broad-scope understanding of what might have happened at the magical beginning of time 13.8 billion years ago. The following two videos will show you what I mean. The first video by Professor Dave is very easy to understand and can help imagine, sort of, what might have happened at the time of the Big Bang. However, although they may be way more appropriate, if you understand the equations used to understand the origin of the Big Bang as described in the next video by Brian Greene, you have my congratulations as you are probably well on your way to becoming a theoretical physicist.

YouTube "*Kinetic Molecular Theory and the ideal Gas Laws / Professor Dave*" (5:11).

YouTube "*Your Daily Equation #30: What Sparked the Big Bang?*" World Science Festival – Brian Greene (37:00).

As for energy, I do not find it easy to understand. But just like Pressure P, Energy E has an equation; $E=mc^2$. This equation simply states that mass, m, is just another form of E and that m becomes equal to E when m is multiplied by a constant, c^2, i.e., which is the speed of light squared. Easy to say but not at all easy to think about, especially when one remembers the magic show of our universe showing up from nowhere 13.8 billion years ago from pure E. So, no, I do not understand E. But I am working on it. And understanding potential energy and the kinetic energy used in coupled reactions will be useful when one grapples with enzymes and how atoms actually go about assembling our fantastic world that's filled with incredible creatures including us. I could go on and on about this subject but it's huge and goes far beyond the scope of this book. However, I can leave you with some wonderful videos that will give you a head's up of what's to come.

***YouTube "The Laws of Thermodynamics, Entropy, and Gibbs Free Energy",* Professor Dave Explains (8:11).**

YouTube *"Steady States and the Michaelis Menten Equation",* Biomolecules / MCAT/ Khan Academy (7:31).

***YouTube "Biochemistry / Michaelis Menten Equation"* Ninja Nerd (22:64).**

YouTube *"75 Chain Reaction Ideas and Inventions",* Sprice Machines (4:25).

YouTube *"Kinetics of Chain Reaction",* (HBr) / Advanced Chemistry (21:36).

***YouTube "How enzymes Work"* (from PDB-101) RCSBpProteinDataBank (4:52).**

I like to go camping and gaze into a fire. I feel the heat coming from the fire and see the smoke and changing colors of the flames. I see the logs as they burn and I think about the amazing chemical reactions that are taking place. All things living in Earth's oxygen atmosphere are thermodynamically unstable, including the parts of trees that are burning in that fire. A burning log is a good example of thermodynamic instability in action. For a log that's a normally slow process of decay which takes many years to complete unless the decay is aided by fire, microbes, or forest termites. But I started the fire in which I now gaze. I sped the instability up with a match and lighter fluid. And once the fire got started, it needed no further help from me. The heat energy expelled by the fire itself creates a temperature greater than the 451 degrees needed to maintain the rapid-demonstration of thermodynamic instability. As I observe the fire, my mind wanders to enjoy how much it explains, including the unstable condition of my very own dissipative self, who I suddenly realize happens to live on this oxygen-rich planet. "Yikes!", My inner voice abruptly awakes from its fire-motivated trance. "Are you trying to tell me I'm as unstable as a log?" "Yup. You've got it. That's exactly right." Now fully awake, I smile, relax and lean back to enjoy the night sky and Andromeda. If I squint and look slightly askance, I can even see the spiral blur of the Andromeda constellation's over-300-billion-star-galaxy with my naked eyes! It never ceases to amaze me. The light from that galaxy takes 2.5 million years to reach my eyes. *I have just time traveled!*

I digress. Back to the log. A log is mainly composed of cellulose. Cellulose is a glucose polymer identical to starch, save for the difference in the linkage of the respective glucose isomers. Cows use bacteria in their rumens that process the cellulose to a cow- useable form. Termites also have such bacteria to process the cellulose to a termite-useable form. We don't harbor such bacteria. So cows can eat hay and termites can eat wood but we can't. But what goes here? Cellulose does not react with oxygen to produce fire until the temperature exceeds 451 degrees Fahrenheit. For that matter, nor does starch. It's surely not that hot in a cow or in me.

To understand this, one needs to understand irreversible thermodynamics, not the thermodynamics that we all first learn about in chemistry class, i.e., reversible-thermodynamics, It's Prigogine's irreversible-thermodynamics, enzyme kinetics, chained enzymatic reactions, first order reactions, mitochondria, RNA, DNA, histones, NAD, ATP, cytochromes, Krebs Cycle, etc., etc. that can lead us to an understanding of the astounding magic that happens in us and all around us each instant of our existence. Every living thing is a dissipative structure that is maintained far from equilibrium by the mystical effects of non-equilibrium, irreversible thermodynamics. This is really exciting stuff and I'm excited to think that this seemingly complex story could be presented in an easy, quick to understand way if it were to be presented in the form of an AWTbook™. Even just one, well- constructed AWTbook™ might be all that one would need to reasonably quickly grasp it all.

For an explanation of Ilya Prigogine's irreversible-thermodynamic dissipative structures, and to just begin to understand that dissipative structures are exactly how you and I show up in existence, check out the following two videos. The first one is a bit graphic but not to worry, the subject is illustrated in cartoons and you are not about to go up in smoke. However, the video might give you pause and an incentive to get busy with your weight-loss program. The second video is an introduction to Prigogine, and his Nobel Prize-Winning work on irreversible thermodynamics.

YouTube, *"A Real Case of Spontaneous Human Combustion"*, The infographic Show (10:07).

YouTube, *"1977 Nobel Prize in Chemistry Awarded Solely to Ilya Prigogine"*, Ichigosatsu McNaughton (4:51).

There is also much to say about atoms and their workings in our current day computers. However, if one were to do an AWTbook™ on that subject one would need to learn a few things about such things as conductors, insulators, semiconductors, transistors, diodes, quantum tunneling, logic gates and the like.

But, there you go, writing an AWTbook™ on the subject might be a great way to learn all about it as one writes. I leave that project for somebody else to tackle and offer the following easily understood introduction to this high tech aspect of Our Mind Boggling Illusionary Universe. So, at my age, I likely won't be writing that AWTbook™, but I definitely think one of my readers should. I can't wait to read your AWTbook™, and be able to say, "Hey Siri" to view your choice of YouTube links, as I read, on my new OLED TV. Okay, I get it. You'll probably use QR codes or some other fancy way to call up links, whatever.

YouTube *"Semiconductors – Physics Inside Transistors and Diodes"*, Physics Videos by Eugene Khutoryansky (13:11).

Epilogue

Nuclei presented at the conclusion of OSAU-1

scandium-45, $^{45}_{21}Sc$ $1s^2 2s^2 2p^6 3s^2 3p^6 4s^3$

Transition metal, rare earth along with yttrium and lanthanoids. Found with uranium deposits. Main use is in aircraft aluminum alloys, especially the Russian Mig-21. But titanium is cheaper and works just as well. A 20%, 20%%, 10%, 20%, 30%; Al, Li, Mg, Sc, Ti alloy is as strong as titanium, light as aluminum and hard as a ceramic.

Check out the part that the atoms with atomic numbers Y-39, Lu-71, Lr-103 play when it comes to Sc-21.

calcium-40, $^{40}_{20}Ca$ $1s^2 2s^2 2p^6 3s^2 3p^6 4s^2$

Calcium is an alkaline earth metal. Ca-40 is the very real, mass-number representation of the calcium nucleus. But you may see calcium written as Ca-20. This is its imaginary, isotopic, atomic-number version sans neutrons. Why write it as C-20? It's easier to remember and easier to relate to other members of the periodic table. To illustrate, I just used above the Sc-21 isotope vs the Sc-45 isotope. But don't forget, the hydrogen nucleus is the only isotope that exists in nature devoid of neutrons.

YouTube "*Alkaline Earth Metals / Properties of Matter*", Chemistry / FuseSchool (5:48).

Calcium is well known to everyone. Much can be said about it, and it's so important much will be said about it in the following video, which illustrates a very complicated way calcium shows up as an important player in Our Self-Assembling Universe, especially when it comes to human endocrinology.

This video also has **Ninja Nerd**, a great classroom teacher who I've found to be good at instructing all things biological.

YouTube "*Endocrinology / Parathyroid gland / Calcitonin", Ninja Nerd (35:03).*

Calcium is also a major atomic ingredient of hydroxyapatite, a key structural component of the human endoskeleton anatomy. This thought led me to a remarkable video on how humans might look in the future.

YouTube "*The Truth Behind the Ideal Human Body in future",* Ridddle (13:13).

Star-Shine through Clouds of Hydrogen Oxide

Here ends OSAU-2, with its Quantum Reality Preface. Our Universe is, indeed, assembling itself. But, before we adjourn, I want to leave you with two more video references. The first one is just stunningly beautiful, wonderfully contemplative and a great example of the type of videography that one can expect to find as a hallmark in AWTbook™ literature. In this very special instance, you will be blessed to witness just a tiny fraction of the vast diversity to be found in the creatures that currently inhabit our planet. It's still hard for me to fathom that all of these beings have been self-assembled **on a daily basis** by the very same atoms that do the same for each of us every day.

The first video also makes it wonderfully obvious that all of us creatures have let our atoms know exactly what we need and want. For example, we humans now benefit from the evolving fitness processes that started on our planet over three billion years ago. Our atoms, somehow, have come to be able to rearrange themselves to marshal in each of us the ~7,000,000,000.000,000,000,000 of them of which we consist to deliver exactly what we need and want. And the atoms have done the same thing for every other creature on our planet. All our wishes and needs for a "happy" survival, when it comes to sex, entertainment, food gathering, defense, etc., have been met. This is incredible is it not? And yet, here we are, and here they are, in all their glory and splendor, right before our very eyes in **8k ULTRA HD.**

So, as you watch this video, I invite you to contemplate Our Universe and the amazing atoms of which we are composed and of which you have now become familiar. I also invite you to think of those atoms feverishly working away, day in and day out, nonstop, for billions of years, in an untiring effort to create and please us creatures. And don't forget to visualize the atoms doing their work using the amazing nanoscale molecular machines they alone invented to accomplish their astounding, creative works. I know, this is a bit anthropomorphic, to say the least. And I grant you there is probably chaos theory and a Superior Being inspired algorithm behind it, but I just can't seem to get over the magic of it all.

YouTube; "Ultimate Wild Animals Collection in 8K ULTRA HD / 8K TV", (16:28).

Having viewed the above video, you might be wondering what life could be like on another planet. The following video, aside from its annoying soundtrack, which I found to be easy to turn off, is a beautiful, well thought-out study on the subject of life elsewhere in Our Universe. The possibilities, as anyone might imagine, are pun-intended, out-of-this-world amazing.

YouTube *"Life Beyond II: The Museum of Alien Life / 4K"*, MELODY SHEEP (38:00).

Our Self-Assembling Universe-2
A Video/Literature AWTbook™
Ends to Exciting Beginnings

About the Author

H. Frank Gaertner is the author or co-author of over 50 scientific papers and patents. In 1958 as a sophomore at Ventura College in California he composed and had performed his work for brass choir. At the University of Arizona in 1962 in a Master's Degree thesis he was first to write about the survival of slime molds in a desert environment. In a 1965 Ph.D. Thesis at Purdue University he was the first to accomplish and write about the *in vitro* synthesis of proteins at elevated temperatures using extracts from thermophilic bacteria (FYI, this was done long before PCR was invented). In 1969 as a Post Doc at UC San Diego he demonstrated the catalytic facilitating properties of multi-enzyme complexes, and in 1979 as a Professor at the University of Tennessee, Oak Ridge National Laboratory Graduate School he discovered the amazing catalytic properties of five enzymes linked together as one polyfunctional enzyme. In 1981 at Salk Institute's SBIA Corp. he isolated the his-*3* gene from *Pichia* yeast for use in SIBIA's new expression-vector. And in 1982 he co-founded Mycogen Corporation to become the company's Director of Molecular Genetics where he headed a team of scientists that made the world's first and only genetically-engineered, organic insecticide. His scientific team also created interferon ARCs, a system for delivering 1000-X, immune-boosting interferon in inexpensive, nano-capsular form. He is frequently in the top ten on ReverbNation for his orchestration of Bach's G# minor Fugue that he renamed *Rivulets* for use in Sally and the Magic River audible book. His other works include *The Amazing Illustrated Word Game Memory Books*, and *Our Self Assembling Universe*. Of all these accomplishments, he regards his most important works to be *Sally's Magic River*, especially its Audible version, *Sally and the Magic River*, and his present AWTbook™, *Our Self Assembling Universe-2.*

A W T References

The following ordered list of abbreviated references with page locations is designed to facilitate one's return to videos and/or particular sections within OSAU-2's new, AWTbook™ presentation. The list can also be used to simplify the creation of one's own video-mix. The author describes, herein, a way to easily make such a personalized list for use as a "just-listen-mix" that can be used in an entertaining, educational, life-extending, walk/jog/run-exercise health-program.

21. Audible.com, The Making of the Atomic----
Bomb, by Richard Rhodes, read by Holter Graham xv
22. Audible.com, Einstein, His Life and Universe
by Walter Isaacson, read by Edward Herrmann xv
23. YouTube, The Story of Energy with Professor Jim Al-Khalili xvii

24. Audible.com, Quantum: A guide for the Perplexed
by Jim Al-Khalili, read by Hugh Kermode xvii
25. Audible.com, "Lifespan" by David Sinclair,
read by David Sinclair xvii
26. YouTube, Your Body's Molecular Machines xviii
27. YouTube, How DNA is Packaged (Advanced) xviii
28. YouTube, Epigenetics Overview xviii
29. YouTube, Life in Miniature: Our Secret War Against Cancer xviii
30. YouTube, David Sinclair New Aging Reversal Approaches 2021 xviii
31. YouTube, Double Lifespan-David Sinclair xviii
32. YouTube, What is an Atom and How Do We Know xix
33. YouTube, Boltzmann's Entropy Equation:
A History from Clausius to Planck xx
34. YouTube, Light and Dark, Both Episodes / Jim Al-Khalili xx
35. YouTube, Is this what Quantum Weirdness looks like? xxiv
36. YouTube, Through the Wormhole-Wave/Particle–Silicon Droplets xxiv
37. YouTube, Do We Have to Accept Quantum Weirdness xxiv
38. YouTube, Sine-Curve and the Unit Circle xxv
39. YouTube, Sine and Cosine from Rotating Vector xxv
40. YouTube, Are Virtual Particles a New Layer of Reality xxv
41. YouTube, Finding the Equation of a Straight Line xxvii
42. Wikipedia.org, Cartesian Coordinates xxviii
43. YouTube, Photoelectric Effect:
History of Einstein's RevolutionaryView of Light xxx
44. YouTube, Demonstrate the Photoelectric Effect xxx

45. YouTube, Blackbody Radiation Oven xxxi
46. YouTube, Quantum Physics-
Part1 Blackbody Radiation - Wien's Displacement Law xxxi
47. YouTube, Black Bodies and Planck Explained xxxi
48. YouTube, Young's Interference Experiment with Single Photons xxxiv
49. YouTube, Why Does Light Really Bend xxxiv
50. YouTube, Why Does Light Slow Down in Water xxxiv
51. YouTube, You Don't Know How Mirrors Work xxxiv
52. Quora.com, Does Light Actually Slow Down in a Medium xxxiv

53. YouTube, How Does a PET Scan Work xxxv
54. YouTube, Atom: The Illusion of Reality xxxv
55. "Hey Siri", The Principles of Quantum Mechanics xxxvi
56. YouTube, Atom: The Illusion of Reality (just repeated to emphasize) xxxvi
57. YouTube, QFT, What is the Universe Really Made of xxxvi
58. YouTube, The Baryogenesis Anomaly: What Happened to all the Antimatter xxxvi
59. YouTube, The Basic Math that Explains Why Atoms are Arranged Like They Are: Pauli Exclusion Principle xxxvii
60. YouTube, What Causes the Pauli Exclusion Principle xxxvii
61. YouTube, Bose Einstein Condensate Coldest Place in the Universe xxxvii
62. YouTube, Quantum computers Explained with Quantum Physics xxxvii
63. YouTube, How Lasers Work – A Complete Guide xxxviii
64. YouTube, Mix – Arvin Ash xxxviii
65. YouTube, Arvin Ash, Are Photons and Electrons Particles or Waves xxxviii
66. YouTube, Make Up Your Mind God! xxxviii
67. YouTube, Can I Explain Feynman Diagrams in 60 Seconds xxxviii
68. YouTube, Richard Feynman Why xxxviii
69. YouTube, The Complete Fun to Imagine with Richard Feynman xxxviii
70. YouTube, Atom: The Illusion of Reality xl
71. YouTube, Introduction to Tensors xl
72. YouTube, This powerful X-ray Laser Can See the Invisible World xl
73. YouTube, Gamma-Ray Laser Moves a Step Closer to Reality xl
74. YouTube, Divergence and Curl. The Language of Maxwell's Equations, Fluid Flow, and More xli
75. YouTube, What the Heck are Planck Units xli
76. YouTube, The Mandelbrot Set – The Only Video You Need to See xli
77. YouTube, Light and Dark 1 of 2 – Jim Al-Khalili xlii
78. YouTube, The Story of Energy with Professor Jim Al-Khalili xlii
79. YouTube, How Information Helps Us Understand the Fabric of Reality, Order and Disorder xlii
80. YouTube, Beyond the Atom, What Really is Everything xlii
81. YouTube, Atom: The Illusion of Reality, Jim Al-Khalili xlii
82. YouTube, The Story of Electricity Full Episode, Jim Al-Khalili xliii
83. Researchgate.net, Chasing the Light: Einstein's Most Famous Thought Experiment xliii

118. YouTube, What's the unique Role of Methane in Climate change 31
119. YouTube, Hydrogen vs Atomic Bomb 32
120. YouTube, Beta Decay 33
121. YouTube, BIG BAY BOOM 2012 35
122. YouTube, All Physics Explained in 15 minutes (worth remembering) 38
123. YouTube, Where Did Dark Matter and Dark Energy Come From 39
124. YouTube, Neil deGrasse Tyson: What is Dark Matter 39
125. YouTube, What is Alpha Radiation 39
126. YouTube, What Makes You ALIVE? Is Life Even REAL 40
127. Wikipedia.org, StellarNucleosynthesis,anAstronomy WikiProject 43
128. YouTube, Visualizing the Planck Length 43
129. YouTube, Steller Nucleosynthesis Explained in 4 minutes 45
130. YouTube, RNA interference (RNAi) 45
131. YouTube, Genome Editing with CRISPR-Cas9 45
132. YouTube , The Most Incredible Recent fully functioning Female Humanoid Robots 2021 46
133. Wikipedia.org, Halo Nucleus 50
134. Physicsforums.com, Shell Structure in a Nucleus 50
135. Astronomicalreturns.com, How the Universe Synthesized JWST's beryllium 56
136. YouTube, Beryllium Unearthed 56
137. YouTube, Avogadro's Number, The Mole, Grams, Atoms, Molar Mass Calculations – Introduction 58
138. YouTube , What's a Lewis Acid and Lewis Base 59
139. Audible.com, Fahrenheit 451 by Ray Bradbury 59
140. YouTube, Reaction of Alkali Metals with Water 61
141. YouTube, Rare Hindenburg Disaster Footage in Color 61
142. YouTube , Introduction to Enzymes and Catalysis / Chemical Processes/ MCAT 62
143. YouTube , Phase Diagrams of Water and CO2 Explained – Chemistry – Melting, Boiling and Critical Point 64
144. YouTube , Vapor Pressure and Boiling 65
145. YouTube , Heat and Temperature 65
146. YouTube , Kinetic Molecular Theory and the ideal Gas Laws 66
147. You Tube, Boltzmann's Constant / Physics 66
148. YouTube , Visualizing the Planck Length. Why is it the Smallest Length in the Universe 66
149. YouTube , Kinetic Molecular Theory and the ideal Gas Laws 67
150. YouTube, Your Daily Equation #30: What Sparked the Big Bang 67

Lightning Source UK Ltd.
Milton Keynes UK
UKHW050837130622
404341UK00003B/12